HOT EARTH HELL!

THE MASTER ALGORITHM

THE QUESTIONS NOBODY'S ASKING!
1. Who's Killing All 'Life' on Planet-Earth?
2. Can Technology 'Save' Us... In Time?
3. What's 'Next' For 10 Billion in 2047?
4. The Secret to Living in Peace and Prosperity ... on Dying Planet?

Paul B Farrell, MRP, JD, PhD

"Where is everybody? Why do we see no signs of intelligence elsewhere in the universe?" Enrico Fermi

"I want to know the mind of God, how he created this world, his thoughts, the rest are details ... The distinction between past, present, and future is a stubbornly persistent illusion." Albert Einstein

"Mankind is in danger of destroying ourselves by our greed and stupidity. Life on Earth is at an ever-increasing risk of being wiped out by a disaster, such as a sudden nuclear war, a genetically engineered virus, or other dangers. The human race has no future if it doesn't go to space." Stephen Hawking

"One of the disturbing facts of history is that so many civilizations collapse," they *"share a sharp curve of decline. Indeed, a society's demise may begin only a decade or two after it reaches its peak population, wealth and power."* In *"so many societies, the elite made decisions that were good for themselves in the short run and ruined themselves and societies in the long run."* They *"managed to insulate themselves from the consequences of their actions ... didn't recognize they were making a mess, until it was too late."* Jared Diamond, Collapse

"From a behavioral economics perspective, we are fallible, easily confused, not that smart, and often irrational, more like Homer Simpson than Superman. From this perspective, it is rather depressing...we are standing in front of a really difficult problem, because people are just designed not to care about this. Global warming will happen in the future ... to other people." Dan Ariely, Predictably Irrational

"Over the last several decades, human activities have so altered the basic chemistry of the seas that they are now experiencing evolution in reverse: a return to the barren primeval waters of hundreds of millions of years ago." Alan Sielen, Scripps Institute of Oceanography

"From an incandescent mass we have originated, and into a frozen mass we shall turn. Merciless is the law of nature, and rapidly and irresistibly we are drawn to our doom." Nikola Tesla

"From the perspective of the source, the world is a majestic harmony of forms pouring into being, exploding and dissolving. But what the swiftly passing creatures experience is a terrible cacophony of battle cries and pain. Joseph Campbell

DECODING THE MASTER ALGORITHM
'HOT EARTH HELL'

The MasterCoders are solving *Hot Earth Hell: The Master Algorithm.* Each focuses on one element of the puzzle. No punches pulled. Brilliant minds. Thoughtful voices—33 MasterCoders summarize thousands given voice during our seventeen years as the #1 columnist on DowJones/MarketWatch, America's leading online financial news organization. MasterCoders voiced future news: what's ahead, the obvious and the denied, the science and myth. Not limited to politics vs science. Nor capitalism vs climate. Not environment vs energy, nor survival vs growth. A comprehensive blueprint, searching for a collective answer to 'The Four Most Important Questions Facing Humans Today: Who Is Killing All Life On Planet-Earth, and Why? Can Technology Really Save The World, in Time? If Not, What's Next for 10 Billion Humans Living in 2047? The Good News: The Solution To Living in Peace & Prosperity … on a Dying Planet?'

TABLE OF CONTENTS: 33 MASTERCODERS

1. "WHO IS REALLY KILLING ALL LIFE ON PLANET-EARTH … AND WHY?"

101. **ASTROPHYSICISTS** *…* Sustainability Bottleneck   25
Enrico Fermi: "Yes, We Really Are All Alone in Our Universe"

102. **SOCIAL INNOVATORS** *…* Culture War Conspiracies   29
Stanford/Yale: "The 'Six-Americas' Civil Wars for Our Souls"

103. **THE BILLIONAIRES … Titanic Rams Noah's Ark!**   33
Big Oil: "Planet-Earth The Titanic, Climate Change The Iceberg"

104. **ANTHROPOLOGISTS … Civilzation Collapsing**   37
Jared Diamond: "How Societies Choose to Fail of Succeed"

105. **THEORETICAL PHYSICISTS … Catastrophic Ending**   41
Stephen Hawking: "Why The Future of Civilization in Space"

106. **SOCIOLOGISTS … Denialism & Our Fear of Death**   45
Ernest Becker: "Wall Street's Insatiable Gorilla"

107. **OPINION POLLSTERS … America's Secret Death Wish**   49
Gallup Inc: "Climate is Not Even in Our Top-20 Fears Today"

2. "CAN NEW TECHNOLOGY REALLY SAVE OUR WORLD ... IN TIME?"

201. **GENETIC BIOLOGISTS ... Gene Pool Madness**   **55**
Richard Dawkins: "Inbreeding Self-Destruct DNA of Capitalism"

202. **SOCIOECONOMISTS ... Inequality Guillotine**   **59**
Joseph Stiglitz: "Gap Worse Than 1929, Even 1789"

203. **CONSERVATIVE IDEALISTS ... Mutant Capitalism**   **63**
Ayn Rand: "Narcissism Killing Adam Smith's Moral Capitalism"

204. **POLITICAL SCIENTISTS ... Global Anarchy**   **67**
Ian Bremmer:" "No World Leaders to Take on Big Challenges"

205. **SILICON VALLEY ... Big Problems, Twitter Brains**   **71**
Jason Polin: "6 Reasons Technology Cannot Save The World"

206. **DISRUPTIVE INNOVATORS ... Creative Destruction**   **75**
Robert Gordon: "Six Headwinds Dropping GDP to 1750 Levels"

207. **ECONOMISTS ... Economics 101, Masters of Illusion**   **79**
Thomas Piketty: "The Bizarre Myth of Perpetual Growth"

208. **PUBLIC ETHICISTS ... Silicon Valley Dinosaurs**   **83**
Peter Diamandis: "Billionaires Avoiding World's Big Problems"

209. **PENTAGON GENERALS ... World War III**   **87**
Michael Klare: "Resource Wars Destroying Human Society"

3. "WHAT'S 'NEXT' FOR 10 BILLION HUMANS AFTER THE 2047 SINGULARITY?"

301. **CLIMATOLOGISTS ... 2047 World's Point-of-No-Return**   **93**
Alex Morales: "Momentum Accelerating, Irreversibly Overheating"

302. **POLITICAL IDEALISTS ... The Shock Doctrine**   **97**
Naomi Klein: "Just Business—Capitalist Profits vs Planet Losses"

303. **EVANGELICAL CHRISTIANS ... The Greatest Hoax**   **101**
U.S. Senator Jim Inhofe: "What If Your 'God' Is Not In Charge?"

304. **<u>REVOLUTIONARIES</u> ... No Plan-B? Revolutions!**   **105**
Chris Hedges: "Science-Denying Capitalists Accelerating Wars"

305. **<u>GEOPHYSICISTS</u> ... Planet's Sixth Species Extinction**
Robert Laughlin: "Hybrids? Recycle? Solar? Earth Doesn't Care!"   **109**

306. **<u>AGRICULTURAL SCIENTISTS</u> ... Peak Hunger**
Jeremy Grantham: "Planet-Earth Cannot Feed 10 Billion in 2050"   **113**

307. **<u>SOVEREIGN LEADERS</u> ... A World Drowning in Debt**
Earth Policy Institute: "$60 Trillion Debt, $75 Trillion GDP"   **117**

308. **<u>INSURANCE INDUSTRY</u> ... Sell Fossils, Buy Insurers!**
Geneva Re-Insurers: "100-Year Disasters Hitting Every 100-Days"   **121**

309. **<u>OCEANOGRAPHERS</u> ... We Are Reversing Evolution**
Sielen/Scripps: "Oceans are World's New Free Garbage Dump"   **125**

310. **<u>RELIGIOUS LEADERS</u> ... Golden Calf of Capitalism**
Pope Francis: "10 Reasons Why 'Creation' is Destroying Us"   **129**

4. "THE SOLUTION TO LIVING IN PEACE & PROSPERITY...ON A DYING PLANET?"

401. **<u>METAPHYSICISTS?</u> ... The 'New Earth' Consciousness**
Eckhart Tolle: "Good News, Living in Peace on a Dying Planet"   **135**

402. **<u>BEHAVIORAL ECONOMISTS</u> ... Zombie Apocalypse**
Dan Ariely: "Yes, Humans are Totally UnPredictably Irrational"   **139**

403. **<u>PHILOSOPHERS</u> ... Morals For Sale, to Highest Bidder!**
Michael Sandel: "Consumerism is Now Killing America's Soul"   **143**

404. **<u>HISTORIANS</u> ... Rise & Fall of Great American Empire**
Niall Ferguson: 'Collapse Hits Suddenly, After Reaching a Peak"   **147**

405. **<u>MARTIAN ENTREPRENEURS</u> ... The Great Escape!**
Elon Musk-Jeff Bezos: "Winning the Lucky Mars Colony Lottery"   **151**

406. **<u>NEW GENDER LEADERS</u> ... Yin = Yang!**
Homans/Rosin: "Global Gender Power is Rapidly Rebalancing"   **155**

407. **<u>FUTURISTS-VISIONARIES</u> ... Star-Crossed Planet-Earth**
Gene Roddenberry: "The Inner Light of Outer Space"   **159**

HOT EARTH HELL!
THE MASTER ALGORITHM

SUMMARY

*The Great AI War: "On a recent Monday morning, Elon Musk
busied himself on Twitter by predicting how World War III would
start. Inspired by news that Vladimir Putin had told Russian students
the country that leads in artificial intelligence will rule the world,
the Tesla and SpaceX CEO declared the global race to dominate
Artificial Intelligence might turn into real war—and that
the first strike could well be launched by an algorithm
rather than a flesh-and-blood leader." Fast Company*

Back at the turn of the 19th Century, automobiles and airplanes were the 'next new thing!' H.G. Wells' *Time Machine* and *War of the Worlds* were popular science fiction. The first computer, the 20-ton ENIAC, was just a dream, four decades in the future. Nickola Tesla was zapped into a new consciousness by three million volts. Albert Einstein wanted "to know the mind of God." For a theoretical physicist known for Theories of Relativity—who believed "the distinction between past, present and future is only an illusion"—wanting to "know the mind of God" was just another way of saying that he was searching for the "Unified Theory of Everything."

Flash forward from 1905 and Einstein's E=MC^2 into the 21st century and today's warp speed world of artificial intelligence, learning machines, big data, social media, wikileaks and fake news. Yet, some things never change. Like Einstein, many are still on the same journey, searching, going beyond. We want to know the "mind of God." That was Einstein's "God Algorithm," now ours. That's what the *New God Algorithm War* is all about, the endless search to know the mind of God, *to help us do the right thing, enjoy the good life, live in peace on a dying planet.*

MASTER ALGORITHMS: EINSTEINS VS ZUCKERBERGS

Peace? The fascinating thing about this war of the master algorithms is that it's actually an accelerating series of fierce battles between opposites, duels, competitions between Einstein's "God Algorithm" on one hand … versus the overwhelming flood of commercial algorithms developed by neuroscientists, mathematicians, computer geeks and hackers, by Silicon Valley's entrepreneurs bankrolled by power players working with Amazon, Apple, Facebook, Google, by Wall Street's quant traders, by the media and every industry where algorithms are their best assets, the key to their power, economic success, earnings, profits, net worth.

Perspective is essential in understanding this perpetual war of the gods. Our human brain evolved by controlling our behavior with the non-math algorithms programmed in our genes. Today however, more and more commercial algorithm creators are controlling our behavior, manipulating our thinking processes, taking over our brain functions with a relentless barrage of mathematical algorithms, in an Orwellian war that humans are losing. In *Sapiens: A Brief History of Humankind,* Yuval Harari hit that nail hard on the head: Humans are "an obsolete algorithm," we are imploding, self-destructing, by some *Grand Design* says physicist Stephen Hawking.

In his classic, *The Hero With a Thousand Faces* Joseph Campbell put it another way; "From the perspective of the source, the world is a majestic harmony of forms pouring into being, exploding, and dissolving. But what the swiftly passing creatures experience is a terrible cacophony of battle cries and pain." In short, any understanding of the Master Algorithm must be conscious of both the "majestic harmony," and also aware we're also living in a virtual, i.e. parallel reality, trapped in a "terrible cacophony of battle cries and pain."

Philosopher and paleontologist Pierre Tielhard de Chardin captures this duality by refocusing our individual perspectives: "We are not human beings having spiritual experiences, we are spiritual beings having human experiences." Today, however, we are drifting out of balance, in fact, we are losing both perspectives, as the human side aggressively reaches for more power in today's wars of the algorithmic gods.

Today, it's no longer enough to know the God's mind, now *we want to become God.* Yet even then, we will be unsatisfied gods, still driven by our human algorithms, the quest for individual wealth, power, fame, where more is never enough where survival is at the core. As Harari put it, "Is there anything more dangerous than dissatisfied and irresponsible gods who don't know what they want?"

COMPUTER SCIENCE REPLACING 'GOD' WITH TERMINATOR

Computer scientists come at this war of the algorithms, as champions for humans in search of a new "unified theory of everything." In *The Master Algorithm: How the Quest for the Ultimate Learning Machine Will Remake Our World,* Professor Pedro Domingos describes it this way:

> "We live in the age of algorithms," they are in "every nook and cranny of civilization, woven into the fabric of everyday life:" Your car, cellphone, house, appliances, toys, banking. "Algorithms schedule flights and then fly the airplanes. Algorithms run factories, trade and route goods, cash the proceeds and keep records. If every algorithm suddenly stopped working, it would be the end of the world as we know it."

In a challenging review, "The Search for the God Algorithm," *Daily Beast's* Nick Romeo warns that Domingos' *Master Algorithm* is based on a conviction that "all knowledge—past, present, and future—can be derived from data by a single, universal learning algorithm," one that will actually "cure cancer, eliminate all jobs" and "Invent everything that can be invented" which he believes is "both possible and imminent."

This very expectation of the imminent creation of this Master Algorithm, makes it clear that Silicon Valley is not searching to "know the mind of God," *they* are "gods," human beings creating math-driven computer tools to help them evolve, replicate and multiply ... and make a nice profit in the process.

Yes, this quest for a Master Algorithm parallels Einstein's earlier search for the elusive "Unified Theory of Everything." And yet, Einstein and more recently Hawking ultimately gave that up that search. However, neither academics like Domingos, Wall Street stock traders nor Silicon Valley's power players in the lucrative technology industry are likely to stop searching for that Master Algorithm any time soon.

Zuckerberg oscillates between these two worlds. When gravitational waves and dark matter hot news last year, he called the waves "one of the biggest discoveries of modern science," one actually predicted by Einstein, "one of my heroes" a century ago in his *Theory of General Relativity.* Zuckerberg also understands the narrow limitations of the human mind: "A squirrel dying in your front yard may be more relevant to your interests right now than people dying in Africa." That's not just his audience, Facebook's global market, he's describing today's culture, politics, economy, worldview, the Zeitgeist, and it is expanding, a perfect metaphor paralleling gravitational waves in physics.

ALL HUMANS SEARCHING FOR A 'MASTER ALGORITHM'

Today there are too few Einsteins searching for the "mind of God." And too many millions of technology geniuses, hackers, unicorns, entrepreneurs and private equity sources chasing the "next new thing," consumer products, apps, gadget, game, toy from some brilliant algorithm creator hellbent on becoming America's next billionaire ... unfortunately, as Hawking and many other physicists warn, chasing short-term economic gains will have disastrous long-term consequences, a catastrophic ending with the "future of human civilization in space."

Yes, here we have two intensely competitive visions of reality, of life, of God—this is the "New God Algorithm Wars," one narrowly focused on Silicon Valley's relentless pursuit of the "next new thing," highly profitable, wealth-building technological innovations.

The other focused on our spiritual nature, aligned with Einstein's desire to "know the Mind of God." This war between algorithms highlights the sharp contrast between the God consciousness of Tielhard de Chardin and the out-of-control materialism, the "unbridled consumerism" that Pope Francis says has taken over the world.

In searching his "God Algorithm," Einstein cautioned us about our limitations, warning us "the human mind is not capable of grasping the Universe." He may be right, yet history tells us humans seem uniquely endowed with intellectual powers, curiosity, boundless energy and an inner roadmap for the journey. Einstein sees our search as a natural process, beginning early:

> "We are like a little child entering a huge library. The walls are covered
> to the ceilings with books in many different tongues. The child knows
> that someone must have written these books. It does not know who or

how. It does not understand the languages in which they are written. But the child notes a definite plan in the arrangement of the books—a mysterious order which it does not comprehend, only dimly suspects."

We are all that child, searching for that "mysterious order," the real Master Algorithm, guided and often manipulated by modern "librarians" who have their own goals and coding agendas.

ALL ALGORITHMS BIASED IN THE WAR OF GOOD VS EVIL

Like so many journalists, *TechCrunch's* Ben Dickson warns against tech's emerging megalomania. In "Why it's so hard to create unbiased artificial intelligence?" he warns us that "the same data used to train machine learning algorithms can teach it to become evil or biased. And like every child, machine learning algorithms tend to pick up the tastes and biases of the persons who rear them," guaranteeing neither benign parenting, ethical leadership, nor a godly algorithm.

Indeed, computer scientists searching for the ultimate 'Master God Algorithm' might first share a lesson in humility from the world of another physicist. In Richard Muller's new, brilliant *Now: The Physics of Time,* he says: "Physics is arguably that tiny subset of reality that is susceptible to mathematics … if an aspect of existence doesn't so yield, we give it a different name: history, political science, ethics, philosophy, poetry … What fraction of what you know that Is *important* is physics? I imagine that even for Einstein that number was tiny."

In his final analysis, Muller goes way beyond the Big Bang, beyond the limited worlds of mathematics, cybernetics and computer science: "Often the most difficult challenge is in asking the right questions; It is hard to know where the next physics revelations will appear," but "advance is unlikely to require complex math or arcane philosophy. Whoever cracks open these problems will likely do so with some very simple examples." In fact, the 33 master codes of today's Master Algorithm wars reflect a search for the right questions about the future of the human species, of civilization, of our home planet.

THE SOLUTION'S HIDDEN IN THE
NEW NON-MATH MASTER ALGORITHM

Gary Zukav, a physicist evolved as a metaphysician, a Green Beret in Vietnam, adds a unique perspective in *The Seat of The Soul,* each of us begin our search *before* we arrive in Einstein's "library," before we are born …

> "Each soul takes upon itself a particular task, it may be the task of raising a family, of communicating ideas through writing, or transforming the consciousness of a community, such as a business community. It may be the task of awakening the awareness of the power of love at the level of nations, or even contributing to the evolution of consciousness at a global level."

Whatever your task, "all the experiences of your life serve to awaken within you the memory of that contract, and to prepare you to fulfill it." As consciousness expands the "majestic harmony" reveals the agreement we made.

Mine became crystal clear in an aha moment, a sudden awareness that the 'task' had already been revealed quite subtly over time in the 1,643 columns posted over seventeen years at DowJones MarketWatch.com, the Wall Street Journal and earlier in MarketWatch's joint venture with CBS and the Financial Times. And after several years was their lead online columnist, analyzing global warming and climate change from a behavioral economics point of view.

My "task" became obvious when looking beyond the limited math coding languages commonly used in computer programming and machine learning algorithms. The MasterCoders for a new "God Algorithms" came in the form of complex non-numeric languages that scientists had dismissed, for lack of measurement and testing, but that were familiar to great minds throughout history from Aristotle and Shakespeare, to Michelangelo, Darwin and many others.

The Master Algorithm's coding integrates many disciplines, the fuzzy logic of quantum mechanics ... chaos theory of mathematicians ... mysterious Zen koans ... earlier technologies first learned as a U.S. Marine Corps radar-computer specialist ... and while publishing *FNX: Future News Index* newsletter for stock traders ... in analyzing J.P. Morgan's use astronomical cycles while a Wall Street investment banker ... researching the Myers-Briggs and Carl Jung's personality types for *The Millionaire Code*... in comparing the 'Prologues in Heaven' in *Goethe's Faust* and the Biblical *Book of Job* for a musical comedy ... then analyzing the coincident of the two Prologues in Heaven with Kurzweil's 'Singularity' in 2047, as our planet passes beyond the point-of-no-return into irreversibly overheating ... while searching for answers to Enrico Fermi 13.7 billion year-old question: "Where is everybody? Why no intelligent life elsewhere in the universe?"

Turns out, I was a just another curious child in Einstein's fascinating "library," actually being mysterious led, as we all are, to rediscover the task I agreed to in an earlier time-space, while searching the "majestic harmony" of our vast universe and negotiating my way through life's "terrible cacophony of battle cries and pain."

QUESTION #1
"WHO IS KILLING ALL LIFE
ON PLANET-EARTH?"

Bill Gates is the world's richest man because of his mastery of computer coding, the mathematical kind, Gates comes at his "God Algorithm" from the same human perspective dominating Domingos' *The Master Algorithm.* In a recent annual report of his charitable foundation', Gates reiterated his ongoing search for an "energy miracle" to solve our global warming problems—new energy cheaper than coal with no pollution—to reverse the trajectory of climate change, to save the planet from the 'catastrophic ending' predicted by Hawking and many other scientists.

Gates' goal is clear: "Zero carbon emissions globally by 2050." His "miracle energy" equation updates one from the earlier work of Japanese energy economist Yoizhi Kaya. Paradoxically, Gates math actually proves that if our civilization doesn't reach zero emissions soon ... our home planet is already dying ... that humans are the cause ... that we are in denial of the obstacles to survival:

Gates' equation is simple: $CO_2 = P \times S \times E \times C$. Translation: Carbon emissions are a function of population growth (P) ... services and resources consumed per capita (S) ... the energy needed to create those services (E) ... and carbon emissions (C) ... Bottom line: Gates equation guarantees a perpetually overheating planet ... that we are in fact already living on a dying planet.

In short, if we really want to save the planet ... we need a radical new approach ... our old thinking, old economics, old technologies, old philanthropists are making things worse, actually killing our planet ... although few of these old-timers will admit they're wrong, as David Owen documents in *The Conundrum: How Scientific Innovation, Increased Efficiency, and Good Intentions Can Make Our Energy and Climate Problems Worse.* Denial becomes a self-fulfilling prophecy.

GATES 'MIRACLE ENERGY' EQUATION: THE PLANET IS DYING

Yes, the world needs more than a new "miracle energy," we do need a new way of thinking, even a new kind of savior. At a recent World Economic Forum, founder Klaus Schwab offered a vision of what's needed, what must emerge, soon. In his new book, *The Fourth Industrial Revolution,* Schwab urges us to "lift humanity into a new collective and moral consciousness based on a shared sense of destiny," to rise above today's materialistic culture.

Schwab gets it: Our world needs a new "consciousness" reversing our struggle, to avoid sinking deeper into myopic political ideologies, economic chaos and cultural conflicts, avoid aligning ourselves with powerful, mysterious dark forces that seems hell-bent on destroying life on Earth, abandoning us as a dying planet.

And yet, "it may already be too late," warns environmental activist Bill McKibben. Still, many pray the "energy miracle" we need will come not as innovative new technologies ... not as fossil fuel alternatives ... not as more investment capital ... nor as increased consumerism ... nor even enlightened politicians ... but in what Schwab identifies as the new "collective and moral consciousness.

ENDTIMES HOAX? NO, EVANGELICALS GOT IT RIGHT!

Paradoxically, conservative evangelicals already know who's in control. In his *Greatest Hoax: How the Global Warming Conspiracy Threatens Your Future,* Big Oil-backed GOP Chairman of the Senate Environment Committee Senator Jim Inhofe apparently "got it," warning us to trust, don't even question:

"God's up there. Humans are arrogant to think we can change what He's doing to the climate." And humans. While scientists analyze natural cycles, conservative evangelicals 'get it' at a gut level, a profound blend of spiritual awareness and material madness that recognizes we are neither cause nor cure.

Moreover, the good senator is not alone; a recent Public Religion Research Institute survey tells us almost half of Americans believe climate disasters are warnings of a biblical "End Times." Similar studies by Gallup and other pollsters tell us that as important as global warming may be, a majority of Americans also believe climate change is not a top-20 national priority.

And while public opinion polls won't *prove* God is to blame for turning Earth into a dying planet, polling results do suggest a lot of Americans are either lulled into inaction or as the recent election clearly proves, they openly oppose climate change policies such as EPA regulations and the U.N. Climate Change Accord.

AMERICANS LOVE PASSING THE BUCK, BUT BLAME 'GOD?'

If evangelicals like Senator Inhofe are right … if God really has absolute control of climate change … if global warming is killing life on our planet … and if humans are powerless to do anything to reverse climate trends … then a helluva lot of people are in fact "blaming" God for the planet's climate problems:

- **BUT WHY NOT JUST BLAME BIG OIL AND FOSSIL FUELS?**
  For failing to plan ahead for the end of The Oil Age? For totally failing to grab the lead in the alternative energy business back forty years ago when Exxon was way ahead in researching alternatives, would have owned that market to the benefit of shareholders, as well as the public. Today these aging fossil fuel companies are still hiding that massive corporate blunder, and now are even making matters worse, as divesting fossil fuels stocks has become so common even Rockefeller heirs doing it, accusing ExxonMobil of "morally reprehensible conduct."

- **BLAME CONSERVATIVE POLITICIANS? AND LIBERAL SCIENTISTS?**
  Why not blame our elected science-denying senators and representatives for blocking EPA rules? Or, blame NASA and the 2,500 liberal scientists on the U.N. Intergovernmental Panel on Climate Change? Scientists are not only convinced humans are the cause, they're also convinced *humans actually have the power to stop global warming.* In the U.N. Paris Accord 197 nations even agreed to save the planet by working together to control carbon emissions and keep global warming to two degrees. Already many nations are hedging and cheating. Meanwhile, scientists warn that new alternative energy technologies can't possibly replace fossil fuels fast enough on a planet of non-renewable scarce resources.

- **WHY NOT BLAME AMERICA'S PROFESSIONAL ECONOMISTS?**
  After all, economists are in bed with their corporate bosses, bankers and the stock market, consistently ignoring the taboo subject of global population growth that is the driver in their perpetual growth myth and our worship of consumerism, the two key forces behind our unsustainable population growth headed for a predicted 10 billion by 2050, steadily depleting scarce, non-renewable resources, accelerating food shortages, creating desert lands, triggering wars, pandemics and natural disasters.

- **EVEN BLAME THE WORLD'S RELIGIOUS LEADERS?** Yes, even Pope Francis. He's got God's ear. In recent years he's been the world's most vocal climate change advocate, even lectured the U.S. Congress about the need to stop global warming and save the planet. And yet afterwards, when he reaffirmed the Catholic Church's outdated birth control dogma, he was adding to the planet's out-of-control population growth. Pollsters now tell us that across the world, he did not change opinions, partisan politics and consumerism trumped pope's message.

Human-caused? Yes, today humans are left holding the bag, get most of the blame for over-heating the planet, from scientists and the media. That's really no surprise.

Way back in 2009, Bill Gates and his billionaires club exposed the problem, agreeing that population was Planet-Earth's #1 ticking time-bomb, accelerating and amplifying all other risks, the primary fuel creating global warming and over-heating the planet. In short, we can't stop the inevitable.

EVEN U.N. IN DENIAL ABOUT POPULATION GROWTH

Population growth has quietly become a taboo topic, in spite of its highly predictably negative impact on economic growth. Population growth has also been minimized or ignored on United Nation initiatives, even though their own demographers predict that 10 billion humans by 2050 is unsustainable.

In the next generation, the planet will add three billion more humans, up from 7.3 billion today. Worse, they will be living in emerging countries with ever-increasing demands for more economic services, higher standards of living, more food, more job opportunities, more wealth creation, more resources ... more of everything.

In the near future, emerging and developing nations will be pushing for more and more short-term economic growth, each driven by their own version of the now universal "American Dream" that has them consuming more and more resources per capita, dumping more and more wastes, as global warming risks accelerate and these emerging economies rapidly mature into developed nations.

This economic trend is unsustainable: Europe and America already consume 32 times more resources per capita and dump 32 times more waste than consumers in emerging markets. Factor that in your algorithms for the 2047 Singularity.

"SUSTAINABLE BOTTLENECKS" KILLING ALL PLANETS?

Physicists are already predicting what's ahead for our planet. Back in 1950 Nobel Physicist Enrico Fermi scanned billions of galaxies and asked: "Where is everybody? Why are there no signs of intelligent life in our universe?" Since then America has spent billions sending out coded messages for other civilizations, scanning galaxies for clues. And still no signs of "intelligent life" anywhere in this vast cosmos of trillions of stars circled by hundreds of millions of potentially habitable planets for the last 13.7 billion years.

Are we really all alone? Astrophysicist Adam Frank offers several possible explanations in a *New York Times* OpEd. One stands out as statistically plausible given our intensely competitive survival-the-fittest economic system: Do "all planets hit a sustainability bottleneck ...none ever make it to the other side?"

Billions of years, billions of galaxies, stars, planets—and yet no civilization has ever become "intelligent" enough to solve their version of the "Master Algorithm," and make it through their sustainability bottleneck to the other side, save itself, and also leave evidence of their survival. Never?

None ever survive? In 13.7 billion years? Hundreds of millions of intelligent life possibilities ... and not one was "intelligent enough" to get to "the other side?" Does every civilization on every planet in every galaxy hit a dead end? Do they all eventually wind up self-destructing when they attempt to break through their unique "sustainability bottleneck?" Is this our fate, repeating a preprogrammed eternal cosmic lifecycle?

SUSTAINABILITY BOTTLENECKS & NATURE'S LIFE-CYCLES

Apparently "intelligence" is not enough, is actually a liability, the real reason no civilization has ever made it through a "sustainability bottleneck." Other civilizations and planets may have evolved, mastered intelligence, also developed megalomania, a God Complex, and begun searching for a new "Master Algorithm" that will save them, help them "invent everything that can be invented" including an essential "energy miracle?" Yet always failing, vanishing, leaving nothing behind.

The sciences already tell us that all life forms are pre-programmed with a 'kill switch.' In human life it's deep in our DNA. All life forms live and die according to their unique cycle, some fast, some slow. But every molecule, every cell, every species has its "cycle of life," of birth and death. Throughout history, every human, every nation, culture and civilization, all planets, stars, galaxies come and go, each to their pre-programmed life cycle. And eventually, all life forms will terminate, exploding in a new big bang or supernova. Or just go quietly, vanishing without a trace. That's the natural cycle of all life forms.

Economist Robin Hanson would call this "The Great Filter," a mysterious "barrier" blocking highly evolved intelligent beings from solving the God Algorithm and making it through their sustainability bottleneck, guaranteeing self-destruction. Then, when a civilization reaches a higher level of intelligence, the kill switch kicks in and they self-destruct leaving no evidence, just a thought transmitted to some future Enrico Fermi on another planet to ask, "Where is everybody?"

BIG BANG BUDDHISM: NO BEGINNING, NO ENDING, NOTHING?

In challenging in the Christian cosmological view, Buddhism and eastern thought leaders generally see past the limited intellectual theorizing and restrictive research of modern science, past the paradoxes and filters of western scientific minds, past the Big Bangs, Black Holes and Supernovas, offering alternative solutions to all the god algorithms, the evolution of species intelligence, and the cosmology of all sustainability bottlenecks.

In the Buddhist cosmology "there is no Creator god," says Brian Schell: "Creation occurs repeatedly throughout time in cycles. In the beginning of each cycle, land forms, in darkness, on the surface of the water. Beings who populated the universe in the previous cycle are reborn." The Dalai Lama clarifies this In *The Universe in a Single Atom, The Convergence of Science and Spirituality:*

"Even with all these profound scientific theories of the origin of the universe, I am left with questions, serious ones: What existed before the Big Bang? Where did the Big Bang come from? What caused it? Why has our planet evolved to support life? What is the relationship between the cosmos and the beings that have evolved within it? ... I am not subject to the professional or ideological constraints of a radically materialistic worldview. In Buddhism the universe is seen as infinite and beginningless, so I am quite happy to venture beyond the Big Bang and speculate about possible states of affairs before it."

Moreover, theoretical physics shares much in common with Buddhism. In *The Grand Design,* while also refuting Newton's belief that "God" must have created the universe because it could not arise out of chaos, Stephen Hawking tells us: "The universe can and will *create itself from nothing* ... Spontaneous creation is the reason there is something rather than nothing, why the universe exists, why we exist. It is not necessary to invoke God to set the universe going."

We also have scientific evidence that from the Big Bang to Black Holes and Supernovas, sooner or later all life does die, all molecules, all cells, all species, every individual, everyone, we all die. In fact, ninety-nine percent of all of Earth's species are *already* extinct.

History expands on this long view: remember, all civilizations, all planets, all-star systems do eventually die, often in sudden catastrophic supernovas after surviving billions of years. The big difference here is that while many religious fundamentalists, cosmologists, climate activists and even scientists focus on big bang beginnings and catastrophic endings, Buddhists, scientists and mastercoders see an infinite recycling of life forms arising over and over out of chaos in many new worlds.

QUESTION #2
"CAN NEW TECHNOLOGY SAVE
OUR WORLD ... IN TIME?"

It is no surprise that physicists like Stephen Hawking and Robert Laughlin agree with Senator Inhofe and other conservative evangelicals. Strange bedfellows indeed, but what they're all really telling us is to forget about the global warming threat, it doesn't matter, climate is changing, overheating the planet, it has momentum, is accelerating, and the odds are, we won't change it.

Why? They see Planet-Earth clearly headed for a "catastrophic ending." In fact, many are telling us to plan ahead for the coming apocalypse. Because we can't stop the inevitable endgame they all see dead ahead.

But is that *God's plan?* Is that the real "Master Algorithm" in action, revealed, a dynamic, evolving system that has no *mathematical* coding? Although they're not directly blaming God, and even if humans really are ultimately the responsible cause, physicists all arrive at the same bottom line—*we are still living on a dying planet.*

Yet, although the warnings are getting progressively louder, so is denialism. Combines, they say a catastrophic ending is dead ahead ... accept your fate ... find a new planet ... plan your exit strategy ... build rockets ... you are living on a dying planet ... admit it ... get off ... Elon Musk and Jeff Bezos' are already working on it ... their SpaceX and Blue Origin rockets are already being tested to do just that, escape ... yes, they're planning a new colony for humans ... on Mars.

But is a cold, desolate Mars colony for a few million humans really a serious solution to global warming on Earth? For our over-heating, over-populated planet? Or is it just an unconscious admission that "something," whether human or divine, really is "killing" life on our planet? And that it really is time to plan ahead for the inevitable endgame, whatever, whenever.

SEARCHING FOR NEW LIFE IN SPACE? FORGET MARS!

A quarter century ago, James Hansen, then director of NASA's Goddard Institute of Space Studies, first warned Congress about the world's global warming threat. Hansen's recently updated research doubles down on those warnings. Climate disasters will accelerate. Hansen warns of a 10-foot rise in sea levels in this 21st century. Meanwhile, the news media and public just yawn.

But coming soon, be forewarned ... expect more Superstorm Sandys ... Manhattan flooding ... Miami underwater ... Predictable 1,000-year dustbowls ... Not long ago insurers measured risk as 100-year floods ... 100-year fires ... 100-year droughts. No more. They're now antiquated seat-of-the-pants benchmarks until recently used by the insurance industry.

The Geneva Association for the Study of Insurance Economics members includes ninety of the world's largest insurers. They have created a new normal. Today the global insurance industry is rapidly shifting to a new business model, using more scientific data and algorithms, now predicting that so-called "100-year disasters" of the past will be hitting the continents every decade or so.

Yes, the insurance industry gets it. No more outdated seat-of-the-pants rules-of-thumb. Nor will they let the self-serving rhetoric of today's science-denying politicians and energy giants dictate tomorrow's insurance rates. They're now using hard climate-science data, not slogans, ideologies and denialism.

2047 BIG CITIES UNBEARABLY HOT! IRREVERSIBLE!

Worse, it turns out that even if humans do hold global warming under the UN-IPCC's new two degrees goal, natural disasters will continue accelerating, the planet will continue over-heating. Why? Climate change "momentum" will just keep heating the oceans and atmosphere, melting glaciers, making the worlds great cities unbearably hot, for living and working.

A recent University of Hawaii study published in *Nature* magazine warns that our world is rapidly becoming one massive hothouse. *Bloomberg News* reviewed the study concluding that not only were "temperatures in New York increasing," but by 2047 most major cities across the planet will pass a "point-of-no-return" into irreversible overheating, *never again returning to historical averages.*

Yes, after 2047 Planet-Earth will just keep automatically heating up for "tens of thousands of years" reports *Climate News Network's* Tim Radford, which will make living and working in major metropolises—including Beijing, Rome, Mumbai, Cairo, London, Rio, Tokyo, New York City—all unbearably hot.

SILICON VALLEY'S NARCISSISTIC 'GOD COMPLEX'

The *New York Times* writes that "in Silicon Valley, there is a sense that tech companies are doing God's work. Software can solve the world's biggest problems, many tech entrepreneurs believe, if only it is put to the right use."

Yes, Silicon Valley's power players are absolutely convinced technology is the solution to saving the world. But will technology's super-optimistic genius innovators really deliver? True, there seems to me a perpetual production line of geniuses minting a steady arsenal of 'unicorns,' start-up tech companies worth over a billion each.

Andrew Leonard of *Salon.com* warns their megalomania is flawed and dangerous: "The tech industry's God complex is getting out of control," quoting Silicon Valley icon Marc Andreessen bragging that "if you live in the San Francisco Bay Area in the early 21st century, it's hard not to feel a special connection to Renaissance Florence." In fact, "start-up entrepreneurs in Palo Alto are just like the Medici Grand Dukes. Except, even better, because they've got superpowers."

Yes "superpowers." Their arrogance and optimism tells them they indeed are the ones who will save the planet with their algorithm-generating powers to "Invent everything that can be invented." However, the *Times* quoted one of the Twitter founder, Evan Williams, who "is less quixotic," admitting that rather than the solution "technology is the direct cause of our biggest problems, global warming, health issues, potential nuclear annihilation."

TWITTER BRAINS SOLVE WORLD'S 'BIG PROBLEMS?'

In a provocative *MIT Technology Review* editorial, "Why We Can't Solve Big Problems," editor-in-chief Jason Pontin challenges Silicon Valley's God Complex: "There is a paucity of real innovations ... technologists have diverted us and enriched themselves with trivial toys."

Silicon Valley recently bragged of more than a hundred "unicorns" in their production pipeline. But their goals are limited, not solutions to the planet's 'Big Problems,' but more focused on making insiders rich solving small problems in social media, consumer gadgets, virtual reality, video games and yet another app. While our planet overheats, slowly pushing America off an economic cliff and into no-growth GDP below one percent predicted by the end of this century.

Unfortunately, if the trajectory continues and Silicon Valley fails to deliver, it'll be too late for "Plan B" in 2047. James Hansen, former NASA guru explained this risk in an open letter to billionaire Warren Buffett: "Recently my oldest grandson, an 11-year old, made the astute observation about climate change: 'Unless we can figure out a time machine that actually works, there will be no way to go back in time to fix it'." Game over, join the Cosmic Sustainability Bottleneck Club.

'AGE OF OIL' ENDING, KILLING FUTURE ECONOMIC GROWTH

Economist Robert Gordon's recent *The Rise and Fall of American Growth* gives us a troubling glimpse into a next Industrial Revolution. A few years ago, Gordon shook up Klaus Schwab's annual World Economic Forum in Davos, Switzerland by presenting his National Bureau of Economic Research (NBER) paper, "Is American Growth Over?" Gordon challenged the prevailing economic mythology that assumes technology will solve the world's 'Big Problems,' thus eventually saving the world from itself:

> "There was virtually no growth before 1750, and there is no guarantee that growth will continue indefinitely ... the rapid progress made over the past 250 years could well turn out to be a unique episode in human history," a collection of "one-time-only inventions." Under that scenario, America's GDP will continue falling. "Now a weak 1.8% average, predicted to sink much further, down into a no-growth 0.2% GDP ... the pre-Industrial Revolution level common across the planet for centuries before 1750."

But what, you ask, about alternative energies from solar, wind, biomass, hydro, tidal? Aren't they an "energy miracle" combo? Can't alternative energies save us?

Unfortunately, by 2050, alternative energies will still be no more than 20% of the world's energy requirements, said Daniel Yergin in *Foreign Policy.* Yergin is the world's leading energy consultant, best-selling author of *The Quest: Energy, Security, and the Remaking of the Modern World.* A realist, he shares neither Silicon Valley's megalomania nor Gates quixotic search for an "energy miracle" that will eventually appear and save the world.

QUESTION #3
"WHAT'S 'NEXT' FOR 10 BILLION HUMANS AFTER THE 2047 SINGULARITY?"

Someone should tell Bill Gates that energy is not the planet's "Big Problem." In an *American Scholar* cover story, "The Earth Doesn't Care If You Drive a Hybrid," Nobel Physicist Robert B. Laughlin warned that "humans have already triggered the sixth great period of species extinction in Earth's history." Yes, humans have already triggered" a catastrophic ending. So, you can forget hybrids, recycling, organic food, solar power, forget clean energy. Not enough. They may make you "feel good," but will never stop the momentum of global warming.

Planet-Earth just doesn't care what we do, to cause climate damage, or to reverse it, it has a "mind" of its own, a part of the "mind" Einstein wanted to "know." About the same time McKibben recast his warning that it's already too late. Now, our last great hope, our best strategy, says McKibben is to keep all the "coal, gas and oil underground." Naturalist Edward O. Wilson even quantified how much that means in his new *Half-Earth: Our Planet's Fight for Life.*

But given the political power of fossil fuel climate deniers and Planet-Earth's out-of-control population, leaving half the planet undeveloped seem highly unlikely, a dangerous "third rail" for politicians facing repeated short-term election campaigns. Especially today with more than a billion cars already above ground moving people around the planet, every day pulling up to the pumps, demanding more and more gas.

Now imagine what happens to all that demand as global population adds another three billion? More people, more cars, more driving, more energy, more pollution, and Planet-Earth keeps getting hotter ... and hotter.

ARE HUMANS PLANET-EARTH'S NEWEST "DINOSAURS?"

Humans may well be the next dinosaurs in the crosshairs of destiny, driven by a selfish survival-the-fittest madness gene, in denial and psychologically incapable of truly comprehending our suicidal drive to self-destruct, opting instead for short-term gratifications. This indictment is obvious in the works of many experts:

- In *THE RACE FOR WHAT'S LEFT: THE GLOBAL SCRAMBLE FOR THE WORLD'S LAST RESOURCES* Michael Klare put it this way: "No matter how much corporate or government officials wish to deny it, there are not nearly enough nonrenewable resources on this planet to perpetually satisfy the growing needs of a ballooning world population." This trend echoes Jeremy Grantham's warnings that it will be "impossible to feed the 10 billion people" that U.N. demographers predict will live on Planet-Earth by 2050, in just one short generation.

- In **COUNTDOWN: OUR LAST, BEST HOPE FOR A FUTURE ON EARTH,** Alan Weisman warns: "Every species in the history of biology that outgrows its resource base suffers a population crash, a crash sometimes fatal to the entire species." In short, while reducing population numbers is not popular with most people and politicians, it's yet another reason even the United Nations is ignoring their own population growth predictions for 2050.

- In **REQUIEM FOR A SPECIES, WHY WE RESIST THE TRUTH ABOUT CLIMATE CHANGE,** Clive Hamilton is also incredibly blunt: Earth will soon "enter a chaotic era lasting thousands of years. Whether human beings would still be a force on the planet, or even survive, is a moot point. One thing seems certain: there will be far fewer of us." George Marshall went further in his *Don't Even Think About It: Why Our Brains Are Wired to Ignore Climate Change.*

- Can we change in time, think different, save our planet, our species? In **HOMO DEUS: A BRIEF HISTORY OF TOMORROW,** Yuval Noah Harari says it's unlikely: "We do not become satisfied by leading a peaceful and prosperous existence. Rather we become satisfied when reality meets our expectations. The bad news is that as conditions improve, expectations balloon." Our demands are addictive, more is never enough.

- So, what is next? As Michael Tennesen details in **THE NEXT SPECIES: THE FUTURE OF EVOLUTION IN THE AFTERMATH OF MAN,** any effort to bring down the accelerating population growth would be "asking humans to do something that no other species has ever done; constrain their numbers voluntarily." Ergo, since population growth is unstoppable, global warming is unstoppable, our sustainability bottleneck is a catastrophic ending.

What if our civilization here on this planet really is the only "intelligent life" left across billions of years, trillions of star systems and hundreds of millions of potentially habitable planets? So far, scientific evidence does suggest that we are in fact all alone. But why us? Are we unique, or merely the *latest* civilization to hit a universally preprogrammed "sustainability bottleneck," and just like all the others before us, we also lack the "intelligence" to figure out the codes, solve our problems, save the world, break through to the other side, and survive.

In fact, "intelligence" may be our fatal flaw, the same fatal flaw in all earlier civilizations on habitable planets. Are we all just too smart for our own good? The latest left alone, living on a dying planet, in denial, self-destructing, victims of our own successes, a limited intelligence. The latest stopped short of the end zone as the clock runs out, unable to break through our "sustainability bottleneck" thanks to the kill switch embedded in our genes … not by some enemy "out there" somewhere.

AFTER "AGE OF HUMANS," A NEW SPECIES? NEW WORLD?

In her classic, *The Sixth Extinction: An Unnatural History,* Elizabeth Kolbert identifies our self-destruct tendency as a "madness gene." The DNA of all humans is instinctively egocentric, we are all capitalists programmed with a survival instinct, a mutation that paradoxically drives us individually and collectively to sabotage not only our own capitalist economy and ideological successes, but to self-destruct our civilization, planet, even ourselves in a suicidal madness. And yes, even setting the stage for the coming "sustainability bottleneck."

Destiny? Or free will? Anthropologist Jared Diamond says the choice is still one that humans can make. Must make. And we are making. In his classic *Collapse: How Societies Choose to Fail or Succeed,* we see a madness gene guiding the collective brains of all species, driving us to risk all, like the mythic Icarus flying too near the Sun, inevitably self-destructing in the relentless pursuit of a dream.

Diamond has warned us so many times that our human brain never learns the lessons of history, that our inbred myopia—short-term survival instinct—makes us perpetually in denial, that we are our own worst enemy in the long-run.

One key example stands out: Centuries ago two million people thrived in the great Mayan civilization. Yet, as with "so many societies, the elite made decisions that were good for themselves in the short run and ruined themselves and societies in the long run."

As a result, the Mayan civilization collapsed "because of a combination of climate change, drought, water-management problems, soil erosion, deforestation." The rulers "managed to insulate themselves from the consequences of their actions." Forests were leveled. Unfortunately, "the kings didn't recognize they were making a mess until it was too late." They were in denial. Sounds familiar?

GRAND DESIGN? GOD PARTICLE? NEW ENERGY MIRACLE?

The catastrophic warnings of physicists Fermi, Hawking, Frank, Laughlin and other experts hint to a far bigger plan for our world, a plan filled with mysteries and contradictions. Einstein believed "the human mind is not capable of grasping the Universe," yet searching for the "Mind of God" was his greatest priority.

So, what are today's Master Algorithm wars really about? At best, experience tells us it is the "still small voice" within each of us. And at the same time, something bigger too. In *The Grand Design,* Stephen Hawking and Leonard Mlodinow hedged their bet, concluding God did not create the Big Bang that triggered our universe, creating life on our planet about 13.7 billion years ago.

More recently, however, when Hawking learned that scientists at Europe's CERN Hadron Collider had finally discovered the elusive "Higgs boson," Hawking again warned that this "God Particle" would eventually trigger a doomsday scenario and "wipe out the universe," echoing his prediction of a "catastrophic ending."

SCIENTIFIC HYPOTHESIS: IS 'GOD' KILLING LIFE ON EARTH?

Our search for the "New God Algorithm" actually began with a challenging hypothesis: A question nobody's asking: "Why is God killing all life on Planet Earth?" And if He's not in charge, who is doing it?

Today, the more accurate way of stating it as a scientific hypothesis would be: Is there a "*God Conspiracy*" killing life on Planet-Earth? What Carl Jung saw under the *Imago Dei,* the underlying "collective unconscious," embedded deep in the DNA driving the brains of all humans, pre-programmed to eventually self-destruct our home planet according to some Grand Plan?

At the time I was working on the script for a musical comedy based on *Goethe's Faust* and its prototype, the Biblical *The Book of Job.* Today everyone's playing the blame game with global warming, passing the buck. Back then the great wisdom in the Book of Job's narrative was obvious in the mysterious, destructive power of God's "Whirlwind" destroying Job's world. Raising great questions, if God really had absolute control of the climate, that global warming is not "human-caused," then you couldn't really "blame" God. Get angry, yes. But then you must also conclude that "God" must have a plan, much like the Dalai Lama's *The Universe in A Single Atom and* Hawking's *The Grand Design.*

We know there are no scientific or mathematically verifiable solutions, although pollsters could easily troll for short-term public opinions, as they did during the recent political campaigns in America and across the world. Such suspicions were later confirmed in a *Washington Post* column: "People are seriously wondering whether God is punishing us with the 2016 election."

So yes, it turns out that that "question no one is asking" really is very much on American minds just below the surface, buried under the climate science denial virus by partisan polls, click-bait bloggers, media hyperbola, political hackers and mean-spirited candidates, all mucking up the collective unconscious.

As Einstein once put it in a moment of self-examination: "The most important question facing humanity—is the universe a friendly place? All people must answer for themselves." And if not friendly, then we must decide what or who is killing life on the planet. If human-caused, then we are our own worst enemies. But what if God is the cause? New questions, more answers.

QUESTION #4
"GOOD NEWS! THE SECRET OF LIVING IN PEACE & PROSPERITY ... ON A DYING PLANET?"

Soon Bill Gates will surrender his quixotic search for "energy miracles." Reality will set in. Then hopefully he will lead the revolution our planet now needs, focusing back on the world's biggest problem—*out-of-control population growth*—which his 2009 Billionaires Club agreed was the world's #1 problem.

Unfortunately, his latest annual foundation editorial suggests he's not yet ready to refocus: "The world is going to need a lot more energy in the coming decades, an increase of 50% or more between 2010 and 2040. But today our biggest sources of energy are also big sources of carbon dioxide, which is causing climate change."

Gates also does admit that alternative energies are "not good enough." While they're "getting cheaper, many developing countries aren't waiting for these tools to become affordable." Instead, "they're building large numbers of coal plants and other fossil-fuel infrastructure ... their people need energy now." More energy now? Wrong, what they really "need" is less people, not more energy.

Deep down Gates surely must know this kind of old thinking, old technologies, old money, old computer science will never solve our world's "Big Problems" as we near global warming's 2047 point-of-no-return. No elusive "energy miracle" will magically appear, stop carbon emissions, save the world from itself, the momentum of temperature will be irreversible. Instead, his outdated algorithm spells disaster. It's time to wake up, create a new global consciousness.

MASTER ALGORITHM: A REVOLUTION IN CONSCIOUSNESS

Yes, we do need a miracle. And fortunately, the World Economic Forum's Klaus Schwab has pointed global leaders to a radical path forward in his new book, *The Fourth Industrial Revolution.* In it Schwab says, "shaping the future" will take more than economics, more than technology—we must "be putting people first and empowering them" lifting "humanity to a new collective and moral consciousness based on a shared sense of destiny."

Schwab gets it: Our civilization needs a "collective and moral consciousness," we must evolve to a higher state of consciousness, take our world beyond the dark materialism, denialism and mutant capitalism that's taken over the Zeitgeist, our global mindset, society's collective consciousness.

Today our world is controlled by an economic system that has lost its moral compass. Until recently scientists dismissed all research into "consciousness" as not "real science." Scientists left consciousness to the philosophers, new age gurus, mystics, psychologists, motivational speakers and religious leaders.

CONSCIOUSNESS: CLIMATE ACTIVISTS VS SCIENCE DENIERS

All that changed in 1990 when Nobel Laureate Francis Crick, co-discoverer of DNA, and neuroscientist Christof Koch published "Toward a Neurobiological Theory of Consciousness." Then in 1994 Crick expanded it into the now ground breaking *The Astonishing Hypothesis: The Scientific Search for the Soul,*

> *"This book is about the mystery of consciousness—how to explain it in scientific terms. I do not suggest a crisp solution to the problem. I wish I could, but at the present time this seems far too difficult. Of course, some philosophers are under the delusion they have already solved the mystery, but to me their explanations do not have the ring of scientific truth. What I have tried to do here is to sketch the general nature of consciousness and to make some tentative suggestions about how to study it experimentally. I am proposing a particular research strategy, not a fully developed theory.*

Crick's hypothesis gave instant credibility to consciousness, moral behavior, ethics and free will as valid subjects for scientific investigation. Our work in behavioral science the past couple decades is an extension of Crick's challenge. Much as Einstein wanted "to know the mind of God," Crick wanted "to know is exactly what is going on in my brain when I see something ... how the rather naive ideas most people have about seeing are largely incorrect," a theme echoed a decade later when Psychologist Daniel Kahneman became a Nobel Economist for his work in behavioral science challenging the assumption of human rationality in decision-making, investment strategies and modern economic theory.

THE EVOLUTION TO A NEW HUMAN CONSCIOUSNESS

After Crick's work on the new science of consciousness, the works of Gary Zukav, Eckhart Tolle and other physicists who earlier expanded into metaphysics took on new meaning. More recently Deepak Chopra M.D. coauthored *How Consciousness Became the Universe: Quantum Physics, Cosmology, Relativity, Evolution, Neuroscience, Parallel Universes* with noted British physicist Sir Roger Penrose. Stephen Hawking and Penrose earlier coauthored *The Nature of Space and Time.*

Zukav's *Dancing Wu Li Masters* was a 1979 work on quantum physics. A decade later he published *The Seat of the Soul,* crediting physicists like Albert Einstein and other scientists as his inspirations because "they reached for something greater than they were able to express directly in their work … they were mystics."

About the same time, another Cambridge physicist Eckhart Tolle published *The Power of Now: A Guide to Spiritual Enlightenment,* and more recently *A New Earth: Awakening to Your Life's Purpose.* In *New Earth* Tolle specifically focused on the link between consciousness, global warming and climate change. In an interview Tolle discussed that link: Every human has the power to tap into their own unique "energy miracle" for the survival of the planet:

> "Yes, it's true that we need to save the planet. But let's not fall into the erroneous thinking that all the solutions are out there somewhere. Because most of the problems—violence, pollution, war, terrorism—all have their origin in human consciousness or unconsciousness. So your primary responsibility is not doing anything outside you; your primary responsibility is your own state of consciousness. And once that is achieved, then whatever you do and whomever you come into contact with get affected by your state ⌐SEP⌐of consciousness."

Tolle's *New Earth* Tolle is clear about consciousness as the prime force in the future of our civilization and planet: "The new earth arises as more and more people discover their main purpose in life is to bring the light of consciousness into this world and so use whatever they do as a vehicle for consciousness."

His message is simple and powerful—step outside the box, stop trying to save the planet with new technologies, new investment strategies, with external solutions. Stop, look within. Endless searching for clean energy alternatives, wishfully thinking about energy miracles "out there," is not productive. Neither is climate science denialism, nor opposition to the UN's Climate Accord. Extremes waste scarce resources, time and money that we can ill-afford to lose if we're serious about breaking through the "bottleneck" and creating a sustainable planet.

SUSTAINABLE CONSCIOUSNESS VS UNSUSTAINABLE PLANET?

Yes, our planet needs a whole new way of thinking, a new kind of energy, new economic ideology, and new kinds of leaders. Most of all we must shift our focus to a "new collective and moral consciousness based on a shared sense of destiny." We must go deep into our souls, fast. Our civilization must shift our paradigm before it's too late, before we pass the point-of-no-return and become what so many already fear is happening—an irreversibly overheating planet.

In the final analysis, a behavioral science solution for the New God Algorithm boils down to a couple of very simple questions: First, will a "new collective and moral consciousness" work as survival strategy, getting us past our sustainability bottleneck, save our planet, our civilization, our global culture?

And secondly, if we cannot save the planet collectively, can a new "sustainable consciousness" help each us *individually* live in peace on a dying planet?

The answers to *both* questions require a personal commitment to living in a new state of consciousness and a lot of work, a tough job in today's deeply troubled, endlessly conflicted world populated and governed by so many who are myopic and narcissistic, in denial of the threat of global warming, and often openly hostile and aggressively profiting from climate change problems, as Naomi Klein detailed in her classic, *The Shock Doctrine: The Rise of Disaster Capitalism.*

Moreover, anyone who is already a science-denier, diehard conservatives, or employed in the fossil fuel industry is likely to reject out-of-hand the idea of raising global consciousness, even though the very real consequences of global warming are accelerating and certain to force even rigid deniers to eventually change, if too late.

In short, getting through our 'sustainability bottleneck' is unlikely to happen voluntarily. We will likely need a massive wakeup call now, a catastrophic event triggering the rapid emergence of a renew personal and collective consciousness, a call that is certain to come but most likely will not be ignited until climate problems get far worse, when it's too late to reverse the increasing temperature momentum.

"POPULATION BOMB" VS THE NEW MASTER ALGORITHM

Together, the MasterCoders of *Hot Earth Hell: The Master Algorithm* offer a clear warning: the odds are stacked against anyone hoping to save the planet from a catastrophic ending. Technology is driven more by their universal god complex than hard science. The warnings have been screaming at us for decades, louder and louder, beginning with the work of pioneers like Rachel Carson and Bill McKibben, culminating in the 2015 U.N. Paris Agreement on Climate Change. And of course, science denialism has exploded into a counter-balance.

Moreover, this conclusion is not simply drawn from all the research cited in the master codes, it's also due to the widespread science denialism imbedded in our genes ... to our ever-increasing consumer demand for more energy, more jobs, more personal wealth, steady economic growth ... to the world's out-of-control, accelerating population growth ... and to our human tendency to delay decisions on Big Problems until it's too late to plan ahead to take action and change course.

Ultimately, getting past our 'sustainability bottleneck' now depends on creating what Klaus Schwab calls a "new collective and moral consciousness" across the "shared destiny" of seven billion people in the United Nation's 197 member nations. Unfortunately, the Master Codes tell us that it's unlikely we will act in time to reverse the accelerating momentum of global warming temperatures.

Why? Here we see scientists, scholars and thought leaders in so many disciplines all warning us the planet's global warming momentum is accelerating so fast that climate changes are locked-in and unstoppable. And if not yet, we are nearing critical mass, will pass the point-of-no-return in the next generation. Why? By 2047 Earth's average temperatures will become a megatrend that will continue on a natural cycle, in an upward trajectory that is irreversible.

GOOD NEWS: THE SECRET TO LIVING ON A DYING PLANET

And secondly, even if no new broad-based collective consciousness emerges in time to save our civilization and planet, there's still good news for individuals. Each of us has the power can make a personal decision and commitment to live in a higher state of consciousness.

Yes, each one of us can live in peace on a dying planet. Seriously, even if global population continues on to ten billion in 2047, the one saving grace might be that so many more individual humans will have the opportunity of choosing to reach for a new consciousness.

Physicist-metaphysician Eckhart Tolle saw this sense of inner peace in Stephen Hawking himself, wrote about it in *A New Earth:* "Although Hawking's body was totally paralyzed and facing what one of his students said was "the most dreadful fate that can befall a human being," still, with the aid of his voice synthesizer, the immobile Hawking still giving thanks for the gift of his life ... "Who could have wished for more."

So no matter what happens as our civilization confronts our planet's sustainability bottleneck ... as global warming gains momentum ... as our temperatures accelerate ... our cities over-heat ... all of us can still live in peace ... in our own personal awareness within a new "collective and moral consciousness," thankful for whatever time we have remaining on the planet, living our lives fully.

We'll probably never see signs of intelligent life anywhere out there in the universe, except in movies. And yet, we can leave signs of our presence, markers that we were here, aware, conscious, alive. We know that "intelligence" is not who we really are, at best it is a beautiful and predictably irrational algorithm created for our brains, so we can function during a relatively short lifetime on the planet.

The truth is, you are a "soul" that never really dies ... that existed before you were born ... that will live on long after your body's journey here is complete ... after you accomplish the task you agreed to perform and committed to another. You are a transcendent soul living in a timeless universe, a spiritual being having human experiences ... aware that past, present and future all exist at once in an eternal now, and like Einstein, you now "know the mind of God," a sustainable consciousness on an unsustainable planet, alive in a new reality, living in peace on a dying planet.

Napoleon Hill's Famous "Preacher's Story"

There's a wonderful old story from Napoleon Hill's all-time classic, *Think & Grow Rich,* about a new preacher who delivered the exact same sermon his first three Sundays in the pulpit. After the third one, nervous eyes darted back and forth across the aisles as parishioners wondered, "Did they hire someone with a mental condition?"

Rumors flew. The deacon was given the uncomfortable task of confronting the new minister. He arranged a private meeting in the parish rectory. He asked the new preacher if he was aware he was repeating himself. "Oh yes," the new minister replied through piercing eyes and a wry smile, "and I'll keep doing it until they get the message."

Readers often tell me they've been reading my books, blogs and columns for many years. My wife however, had chided me for so often punctuating a point with a terse, "Get it?" But like Napoleon Hill's preacher, I also know that repeating themes, facts, quotes and messages is often necessary, to anchor what really counts … to jolt the audience awake in today's overwhelming, noisy, distracting culture … to get people really thinking in a world where truth, love and god are all too often casual clichés and over-valued commodities.

* * * * * * *

Researching 1,643 MasterCoders

The following 33 MasterCoders were originally published as columns on DowJones-MarketWatch between 2007 and 2015. They have been edited, revised and combined as a master algorithm that collectively is not only more relevant today, but an accurate predictor of the future of Planet-Earth. The individual columns remain part of a larger archive of 1,643 columns and nine books published during my seventeen-year tenure, first as an editor then as MarketWatch's #1 columnist focusing on the economics, politics and morality driving climate change and global warming. After leaving DowJones, the momentum behind these individual MasterCoder reviews clicked, the bigger picture became obvious, a story that needed telling, when suddenly—like one of those aha moments that hit you while meditating on a Zen koan—*Hot Earth Hell: The Master Algorithm* became the obvious focus of this my tenth book. This book has been edited so many times over the years, right up to the last minute before publication, so please feel free to let us know if you see the glitches or have any comments. We'd often get thousands of comments when the columns were originally posted on DowJones MarketWatch! Paul@PaulBFarrell.com.

QUESTION #1:

"WHO'S REALLY KILLING ALL 'LIFE' ON PLANET-EARTH?"

SUSTAINABILITY
BOTTLENECK

"Our climb up the ladder of technological sophistication comes with a heavy price. From climate change to resource depletion, our evolution into a globe-spanning industrial culture is forcing us through the narrow bottleneck of a sustainability crisis. Maybe not everyone, maybe no one, makes it to the other side."
Adam Frank, About Time: Cosmology and
Culture at the Twilight of the Big Bang

NO PLANET GETS PAST THE 'SUSTAINABILITY BOTTLENECK'

Physicists probe deep space, back and forward across billions of years of exciting cosmic history, light-years across black holes and ancient galaxies, collecting endless data, scientific, mystical and mythical. If there ever was anybody out there, they not only vanished, they disappeared without a trace ... not one intelligent civilization left anything behind.

"Where is everybody?" famously asked Nobel physicist Enrico Fermi in 1950. "Why do we see no signs of intelligence elsewhere in the universe?" We're still asking.

Our human race has only been around for maybe two million years, a fraction of a second in cosmic time on a planet that's over thirteen billion years old. And although we've spent billions of dollars on sending coded messages into space and listening on satellites for signs of life, it seems nobody's out there, anywhere.

The Search for Extraterrestrial Intelligence SETI organization says there are a billion potentially habitable planets like Earth and probably 100,000 intelligent civilizations in our galaxy alone. And yet SETI has never ever picked up any signals, no signs of intelligent life, nothing, from not one of them, ever. Never! Are we really all alone? Never contacted? No evidence? Why?

NO SIGNS OF INTELLIGENT LIFE ... WHERE IS EVERYBODY?

For human civilization, our greatest problem, our greatest challenge and by far, our most exciting opportunity has been defined by astrophysicists. It's not drones delivering UPS packages or Amazon books ... not a brief expensive tour into weightlessness aboard Branson's Virgin Galactic ... a trip with Musk's SpaceX or Bezos Blue Origin Mars Colony ventures ... nor is it Planetary Resources potentially lucrative private equity venture into asteroid mining.

Physicists tell us space is the greatest opportunity for our highly evolved capitalist civilization. That's the vision of future for the world's leading astrophysicists, environmental activists like Bill McKibben's 350.org and the Grantham Research Institute for Climate Change It is constantly challenged by Big Oil capitalists and the 67 billionaires who already own half the world's assets, as well as Silicon Valley's innovation genius machine.

If these physicists are right, time is urgent. We have no choice. We must solve the real mystery of "space, the final frontier." And fast. Here's how four of the world's leading physicists see great opportunities and why we must act now to avoid the future of human civilization on Earth simply repeating past cycles on billions of universes, stars and planets over billions of years since the Big Bang, leaving no evidence of their existence. Consider these four:

- **FERMI ... NO OTHER CIVILIZATIONS SURVIVED IN 13.7 BILLION YEARS?** Nobel Physicist Enrico Fermi, the "father of the atomic bomb," is also known for the Fermi Paradox, the infinite gap between a belief that there *must* be or have been other civilizations somewhere "out there" in the billions and billions of galaxies ... versus the total lack of evidence. High probabilities, no evidence. "Where is everybody," Fermi asked, why "no signs of intelligent life in the universe?"

- **FRANK ... CAN HUMANS CREATE A SUSTAINABLE PLANET, IN TIME?** Fermi's Paradox prompted astrophysicist Adam Frank, author of *About Time: Cosmology and Culture at the Twilight of the Big Bang* to add rhetorically in a *New York Times* Op-Ed: "Do all planets hit a sustainability bottleneck and none ever make it to other side?" Why ask? Apparently "civilization inevitably leads to catastrophic planetary change." Frank says climate is "an existential challenge for civilization." He's still optimistic: "Human beings are remarkably clever and inventive... I'm sure that we are capable of dealing with the climate-related challenges we face."

- **LAUGHLIN ... IS OUR HUMAN SPECIES PRE-PROGRAMMED FOR EXTINCTION?** Writing in the *American Scholar* journal, Nobel physicist Robert B. Laughlin warns that "Earth doesn't care if you drive a hybrid!" Or recycle, eat organic food, or live in a greenhouse powered by solar energy, or join Greenpeace, fight Exxon Mobil to save natural resources under the Arctic. Since the Big Bang   13.7 billion years ago, Earth has been operating on its own timetable, doesn't really care if you pollute the atmosphere with CO_2. Why? Every now and then Earth just wipes the planet's ecosystem clean, either a new ice age or burning hell. Physicists are now certain that human behavior is the cause of the coming sixth great species extinction.

- **HAWKING ... MUST HUMANS BUILD A WHOLE NEW CIVILIZATION IN SPACE?** Physicist Steven Hawking sees a "catastrophic ending" for Planet Earth in the not-too-distance future. As a result, "the long-term future of the human race must be in space," says Hawking. So, it's important that we start planning for a new life after Planet Earth: "It will be difficult enough to avoid disaster on Planet Earth in the next hundred years, let alone the next thousand," before the sixth great extinction wipes out our civilization before our capitalist innovators and space explorers can find a new home for human civilization.

But nothing stops adventurous souls from searching, no matter what the obstacles and how long the odds. Back in 1969 President John F. Kennedy's science adviser has this to say after the landing on the Moon: "It would be a mistake to commit $100 billion to a manned Mars Landing when we have problems getting from Boston to New York City." Are the benefits better today, the costs, the odds worth it?

KILLER COSTS OF THE 'SUSTAINABLE BOTTLENECK'

MIT Technology Review's editor-in-chief Jason Pontin also anticipated Silicon Valley's dilemma in his provocative article, "Why We Can't Solve Big Problems," facing the issues defined by great minds like Fermi, Frank, Laughlin and Hawking.

Polin's answer: Too many limitations, like the fact that our problems are often misdiagnosed. They're not technological, but rather constrained by economics (like the estimated $60 trillion cost of stopping global warming in a $75 trillion GDP world). Or the fact that there are bigger public policy demands here on Earth (the seemingly insurmountable issues of poverty, inequality and human rights). Or the endless divisive clashes of domestic and international political ideologies, with little prospect of compromise.

SHORT-TERM THINKING KILLING LONG-TERM SURVIVAL

That gets us to the core underlying problem: Clayton Christensen, one of Silicon Valley's great minds, foremost authority on capitalism, and inventor of the new science of "disruptive innovation," believes capitalism is broken. Needs a fix. And last year he made clear the solution in his article, "The Capitalist Dilemma," in the *Harvard Business Review.* Listen closely:

"Sixty months after the 2008 recession ended, the economy was still sputtering, producing disappointing growth and job numbers. Corporations seemed stuck: Despite low interest rates, they were sitting on massive piles of cash and failing to invest in new initiatives."

The core issue is "that investments in different types of innovation have different effects on growth but are all evaluated using the same (flawed) metrics," namely *short-term* investment criteria like closing prices, quarterly earnings, annual bonuses, which results in capitalism's failure to invest in the big long-term opportunities.

Unfortunately, Wall Street, Silicon Valley and indeed the entire American capitalist machine is increasingly more—not less—focused on the short-term, on things like closing prices, quarterly earnings, annual bonuses making it less and less likely they will shift and take on the long-term solutions necessary to take advantage of the opportunities in the problems, challenges and opportunities in the questions raised by astrophysicists Fermi, Frank, Laughlin and Hawking.

CAPITALISTS: TOMORROW UNCERTAIN, LIVE FOR TODAY!

And unless we can solve this Capitalist's Dilemma that Christensen clearly articulates—*by shifting our economic focus from short-term profits to the long-term vision*—humans will have trouble taking advantage of opportunities inherent in these several key questions raised by physicists: Will the intelligence of Planet-Earth help humans become the first to break through and survive a sustainability bottleneck? Who's killing life in Earth? Can technology save us by creating a sustainable planet? If not, what's next for the human species? Can a human live in peace on a dying planet?

The details are mindboggling: Do we have enough capital, innovative talent, and time to plan ahead, to relocate billions of humans in space, somewhere out there? ... Or maybe enough time just to relocate a few human survivors to breed and build a new civilization? ... Or is it too late? ... Is civilization headed for a catastrophic planetary change, and there's nothing we can do about it?

So many unanswerable questions. Is that what's really behind Fermi's ultimate question: "Where is everybody? Why are there no signs of intelligent life in the universe?" Or just no signs that we can see? Apparently, intelligent life must be out there, somewhere ... but apparently, they also hit a sustainability bottleneck ... and none left any evidence behind ... ever ... should we?

5.21.15

CULTURE WAR
CONSPIRACIES

"Climate Science as a Culture War: The public debate around climate science is no longer about science. Physical scientists may set the parameters for understanding the technical aspects, but they do not have the final word on whether society accepts or even understands their conclusions. We must acknowledge that the debate over climate change, like almost all environmental issues, is a debate over culture, worldviews, and ideology." Stanford Social Innovation Review

CLIMATE HOAX? BIG OIL, GOP, EVEN 'GOD' WARNS US!

Yes, climate science really is a hoax. Not because the problem is obvious or scientific. Why? Because there is no solution. Talk's cheap. But there are no real solutions. Worse, no political will. Too much science. Too little consensus. No real action. Nothing. America's lost its moral compass. Sits on the sidelines. Lost the can-do spirit that inspired me as a U.S. Marine, made it a great nation.

And the clock just keeps ticking: Thirty million Americans simply do not trust scientists warning of a "97% certainty" that humans are causing global warming. But they do trust Big Oil, the GOP, even God. They honestly believe climate science is a dangerous fear-mongering liberal conspiracy. Listen, we'll explain:

When Florida Sen. Marco Rubio joined the deniers, he said: "I think all science deserves skepticism." And in a debate between four candidates in the GOP primary for North Carolina governor all denied climate change was manmade, agreeing with the current governor, a former long-term Duke Power executive.

Yes, their party's position is clear, mapped out by Oklahoma Sen. James Inhofe in *The Greatest Hoax: How the Global Warming Conspiracy Threatens Your Future.* But in an effort to question Inhofe's motivation, a ClimateProgress.org review noted that over the years Inhofe has received "$1,352,523 in campaign contributions from the oil and gas industry, including $90,950 from Koch Industries."

Also challenged, Inhofe's reliance on divine guidance. Inhofe said "God's still up there. The arrogance of people to think that we, human beings, would be able to change what He is doing in the climate is to me outrageous." But before you dismiss Inhofe, or any other science deniers for their strong religious convictions, remember the great Christian King Canute of Denmark who sat on his throne at water's edge, to prove the futility of commanding the tide to stop.

WHO CAN YOU TRUST? BIG OIL? POLITICIANS? YALE? GOD?

Environmental economist Bill McKibben wrote in *Foreign Policy*, it may "already be too late" to stop the impact of our climate change. New evidence keeps piling up ... a new Lloyds of London report urging Munich Re and other insurers to factor climate risks in their pricing models ... Pentagon generals are warning climate change is now a threat to national security ... plus a new report by S&P Rating Services warns of sovereign nation credit rating downgrades spreading globally as the rising costs of climate disasters put pressures on economic growth and progress.

Nobel physicist Robert Laughlin hammered home a similar point in his *American Scholar* cover story, "The Earth Doesn't Care If You Drive a Hybrid," warning that humans are not only causing today's climate change disasters, but are fueling Earth's "Sixth Great Species Extinction," which may result in our civilization disappearing like dinosaurs. Drive a hybrid? Recycle? Eat organic? Solar energy? All mere Band-Aid solutions. Science, technology isn't the problem.

Inhofe is not alone in believing God's in charge, humans can't turn back the tide. The fact is, climate science really is a big "hoax" to millions of Americans. The *Stanford Social Innovation Review* diagnosed this trend in "Climate Science as a Culture War" a few years ago. Their research is clear: "The public debate around climate science is no longer about science, it's about values, culture and ideology." Yet we keep trying to convince them with more science. They don't care.

MULTIPLE PERSONALITY DISORDER: WAR ON OUR SOUL

In its research, *Stanford Review* build on the Yale University's "Six Americas" opinion research study, updated annually, which concludes that big "majorities believe global warming will harm future generations of people and plant and animal species." But "four in ten say they feel helpless, disgusted or sad when thinking about global warming."

Yes, scientists may be certain, but the American soul is deeply divided, at war with itself ... psychologists would diagnosis this a multiple personality disorder ... trapped in multiple mental conflicts between the "Six Americas" ... six separate personalities that don't get along ... wars accelerating with each new climate disaster... while distrust of science gets darker ... while the political, ideological, cultural gap between these "Six Americas" widens ... while the hoax metastasizes into an even bigger hoax further dividing Americans ... the clock keeps ticking.

No solution. Just more research: Endless warnings from scientists at UN-IPCC, US-NCA, insurers, generals, rating agencies. Just louder warnings of a "95% certainty" the danger is bigger. But no solution, no consensus, no political will. Why? The superrich, powerful, conservative 11% segment of "Six Americas" fights hardest, spends more, is more aggressive, nearly doubling from six percent five years ago. Believe passionately. Yes, 11%, 30 million science-deniers who are fighting a hoax.

This "Dismissive America" is one of Yale's "Six Americas:" They are certain "climate change is not happening ... does not warrant a national response." Many are "high-income, well-educated white men ... very conservative Republicans ... civically active and hold strong religious beliefs ... likely to be evangelical Christian ... strongly endorse individualistic values and oppose most forms of government intervention." Who are they?

Yale doesn't name names, but the profile may well fit Inhofe's political donors.

What's next if the "Culture War" continues? New surveys show 76% of Americans at least passively agree with Inhofe, say climate change is not a top national priority. And that, unfortunately, suggests predictions made by Australian Public Ethics Professor Clive Hamilton in his "Requiem for a Species: Why We Resist the Truth about Climate Change," may prove all too accurate: Soon Earth will "enter a chaotic era lasting thousands of years. Whether human beings would still be a force on the planet, or even survive ... one thing seems certain: there will be far fewer of us." Science is no solution to our culture war.

OUR HUMAN BRAIN IS WIRED FOR DENIALISM

Folks, we just keep asking the wrong questions. We have no solutions. And no will to act. Yes, Stanford, Yale, McKibben, and millions of others are among the other five "Six Americas" who believe they have a solution. More science. They're wrong. Just Band-Aids. Silicon Valley capitalists love searching for the next scientific solution, next Big Thing, next profitable IPO.

But nobody has any real solutions to the real problems: A deeply dividing "Culture War," as climate disasters accelerate ... lost in magical thinking ... six multiple personalities locked in a costly, deadly battle for the Soul of America ... like lost characters in Beckett's classic, "Waiting for Godot," endlessly searching for answers in the wrong places, trapped in Sartre's existential "No Exit" hell.

Yes, so many wealthy and well-intentioned great minds sure keep trying: The UN-IPCC team of 2,000 scientists now on their fifth assessment since 1988, the 500 scientists at the U.S. National Climate Assessment, NASA's Jim Hansen, Bill McKibben and his 350.org global network of climate activists, who were arrested at the White House protesting the Keystone XL pipeline, plus the new "Risky Business" team of former New York Mayor Michael Bloomberg, Hank Paulson and tech billionaire Tom Steyner ... all perfectly rational Americans ... the list goes on.

CAN WE STOP ASKING BAD QUESTIONS ... ACT IN TIME?

Yes, we just keep asking the wrong questions: Yes, the problem exists, it's so painfully obvious ... But do we need more research? ... No, we need consensus, political will, action ... We need to stop avoiding the taboo issues ... Ask the right questions ... How do we get a consensus with "Six Americas" so divided? ... Can anyone stop the inevitable? The estimated 50% increase in global carbon emissions by 2050? ... How in a world of 190 nations with little trust, who don't like America telling them to cut emissions that will slow their economic growth ... And where's the new technology for all nations, not just Silicon Valley's next profitable Next Big Thing IPO? ... And can we ever stop in time?

Time to refocus America, unite all "Six Americas." Face the real problems. Admit throwing more science at science-deniers isn't working. Admit climate science really is a real "hoax" to millions of politicians, energy billionaires, evangelicals.

The "Dismissive" just aren't buying the science. In fact, more science, more technology just toughens, intensifies, accelerates the resolve and resistance of America's 30 million "Dismissives." Makes them stronger. Widens the cultural gap. Time to refocus, even downplay science? Clock's ticking. Emissions won't stop.

Does America really need a new consensus, new political will? Or are indeed destined to follow the dinosaurs into extinction. Remember Ben Franklin's warning at the signing of the Declaration of Independence on July 4, 1776: "We must, indeed, all hang together, or assuredly we shall all hang separately."

5.27.14

TITANIC!

RAMS NOAH'S ARK

"Mr. Buffett, in your last annual report you conclude that continued inaction on climate change 'is foolhardy, call this Noah's Law: If an Ark may be essential for survival, begin building it today.' Recently my oldest grandson, as an 11-year old, made the astute observation about climate change: 'Unless we can figure out a time machine that actually works, there will be no way to go back in time to fix it.' Mr. Buffett, young people will not get a 'do-over.' They urgently need us to start building your Ark." James E. Hansen, former Director of NASA Goddard Institute for Space Studies

PLANET-EARTH, THE NEW TITANIC ... CLIMATE THE ICEBERG

The world is sinking. And the band keeps playing: On the Titanic, first violinist, Big Oil billionaires. For them, capitalism is the solution to everything. Second chair, world's moral authority, Pope Francis warning that capitalism is the "root cause of the world's problems." No harmony there. And playing that haunting flute solo, Mother Nature. She doesn't care what the fossil fuel giants do, doesn't care what Francis says.

Abandon ship? Surrender to the siren song of the climate science deniers? Maybe. Pope Francis's tune is not seductive enough to win the night. True, the pope is the world's moral authority. But morality will never trump the Koch brothers' $100 billion bankroll, certainly not in time to avoid the icebergs. The Koch empire's pledged $889 million to win 2016 election for fossil fuel lobbyists, bolstered by "No Climate Tax Pledges" that conservative members in Congress must sign to get Koch campaign cash. Yes, money talks.

Get it? Money always trumps morality in today's out-of-control capitalist world. Democracy, our moral compass, and America's future are all for sale to the highest bidder. Everyone and everything has a price, especially in Washington.

Why can't Francis, the world's moral conscience, lead a resistance movement against Big Oil and the Koch Empire? Save the world? True, he does lead a powerful army of 1.2 billion Catholics worldwide ... and, yes, his papal encyclicals carry great weight, making official his position that climate change and global warming are indeed manmade ... that capitalism is the "root cause" of all the world's deteriorating physical and social environment ... that humans are killing their own home planet.

Pope Francis has been traveling the world warning that capitalism is the arch-enemy of Planet-Earth: In capitalism, the "worship of the ancient golden calf has returned in a new and ruthless guise in the idolatry of money ... lacking a truly human purpose" ... our "constant assaults on the natural environment" are "the result of unbridled consumerism" ... having "serious consequences for the world economy" ... capitalism is morally destructive of the world's soul and your own soul ... capitalism is suicidal, will inevitably self-destruct, will take the planet down with it.

SCIENCE DENIERS: DESTROYING 'GOD'S CREATION IS A SIN'

The message Pope Francis delivered in his encyclical on global warming is that our destruction of creation is a sin. This September, when he addresses a joint session of Congress in Washington and the U.N. General Assembly in New York, his message will resonate with world leaders, as it will during the Paris negotiations of the U.N. Intergovernmental Panel on Climate Change just before Christmas.

In labeling environmental degradation a sin, the pope is branding one helluva lot of people "sinners": Big Oil, the Koch brothers, the GOP and the 144 senators and elected members of the House who are Catholics — many of them on record as admitting they are climate-science deniers.

Get it? Pope Francis has branded them "sinners" for failing to lead ... for denying humans are the cause of climate change and global warming ... for rejecting scientific reality ... for failing in their sworn duty to face the fact that we are putting deadly carbon emissions in the atmosphere.

No wonder there's high anxiety on Capitol Hill among those climate-science deniers in Congress. Imagine them facing this moral and financial dilemma when Frances is preaching to them face-to-face at the joint session of Congress in September, in effect forcing them to choose between being a sinner in the eyes of God, and getting Koch money for re-election. High anxiety.

NO-WIN WAR FOR BILLIONS OF HUMANS WORLDWIDE

Yes, the Koch brothers are hoping their $889 million commitment to the 2016 presidential election cycle, and then their army of "No Climate Tax" signatories in Congress, will trump Pope Francis and win the war. The truth is, the Kochs really don't care who wins: Their real goal is just to keep big money flowing uninterrupted into Big Oil coffers, no matter who's president.

On the other side of the battlefield, of course, many are hoping Pope Francis will change the minds of America's GOP-dominated Congress. And then rally 200 world leaders at the U.N. General Assembly.

Warning: Both sides are wrong. This is a no-win war, a dead-end; deny it all you want. Mother Nature has her own plans: Global warming is inevitable, and climate change is destroying our world.

True, the Koch money and the deniers are accelerating the end game. But nobody, not Pope Francis, not all the world leaders, and not the 2,500 scientists in Paris at the fall meeting of the U.N. Intergovernmental Panel on Climate Science can stop Mother Nature's inevitable cycles.

Here's why: Five years ago, Nobel-winning physicist Robert Laughlin of Stanford wrote a chilling piece in the *American Scholar* journal: "The Earth," he wrote, "doesn't care if you drive a hybrid." And it's also clear Mother Earth doesn't care if we recycle, eat organic food, or live in house powered by solar energy.

EVEN U.N. PARIS AGREEMENT WON'T STOP OVERHEATING

Joe Romm published an editorial in *ClimateProgress.org* titled: "Of course Paris climate talks won't keep global warming below the dangerous 2 degrees limit." Romm's brilliant piece is yet another warning on par with Laughlin's and many others: a grim reminder that Mother Nature won't care even if every climate science denier magically do an about-face and start funding Bill McKibben's 350.org, grass-roots climate-change organization.

Or if they donated $100 million to RiskyBusiness.org, the new early-warning system for corporate businesses whose members include Michael Bloomberg, Hank Paulson, tech billionaire Tom Steyer and other national figures like Robert Rubin, George Shultz, Olympia Snowe and even Cargill's boss, Greg Page. Mother Earth will just keep warming, soon the momentum will pass the point of no return.

OUR CASABLANCA MOMENT: "WE'LL ALWAYS HAVE PARIS!"

Romm's warnings remind us of the Humphrey Bogart classic "Casablanca." Recall his dramatic goodbye to the love of his life, "we'll always have Paris," just before she boards a small plane with her husband, a World War II resistance leader returning to the fight. Pope Francis is a similar symbol of eternal hope, against all odds.

The UN-IPCC's 2015 Paris agreement in carried such utopian hope, while Romm is one of America's best and most realistic journalists in the climate-change arena. Here's his take, beginning with some esoteric scientific comments on charts from the Climate Interactive and MIT:

"The world has been headed beyond a catastrophic pathway, which would take us ultimately to 6°C warming or more. We have been ignoring climate scientists for so long, more than a quarter-century, that there was never a possibility that one agreement could change our emissions pathway so sharply."

Never? Yes, never. We're trapped in a catastrophic no-win war: "Paris will 'not get us onto the 2°C pathway,' as Christiana Figueres, the top UN climate official said. That wouldn't be possible "without every major player agreeing to specific and serious post-2030 cuts, an outcome that was never on the table." Yes never. Why? The General Assembly of 198 nations includes aggressive oil-revenue giants like Russia who will never cooperate.

Capitalism in action. Profits and wealth accumulation always trump morality — and environmental survival. The fact is that "there are simply too many major political leaders in this country and in other key countries" who are climate-denying capitalists and "simply do not understand how dire the situation is, in part because of the most well-funded disinformation campaign in history, coupled with an equally well-funded lobbying campaign against climate action."

TITANIC CRASHES ICEBERG, SINKING PARIS LOVE STORY

Then comes the hope-filled "Casablanca" liftoff moment: "Certainly Paris should reaffirm that the goal of the ongoing process is to get onto the 2°C path. Failure to ultimately avert catastrophic climate change would rightly be judged by future generations as the greatest failure in the history of humanity, though that failure will be tied to all of us, not just those directly involved in the U.N. negotiation process."

Yes, "greatest failure in the history of humanity," as climate sinks the Titanic. With you, me and the other billions, all sharing the blame. Not just Big Oil. Not just billionaires like the Koch Bros. We're all "sinners" says Pope Francis. Soon it will be all over. Remember Stephen Hawking's prediction, our planet only has two centuries before a catastrophic ending, before we become a hot new Venus, or the next cool Mars—he says our future is in space!

2.17.15

CIVILIZATION
COLLAPSING

"One of the disturbing facts of history is that so many civilizations collapse." They "share a sharp curve of decline. Indeed, a society's demise may begin only a decade or two after it reaches its peak population, wealth and power." Jared Diamond, Collapse: How Societies Choose to Fail or Succeed

BILL GATES BILLIONAIRES: POPULATION IS OUR #1 PROBLEM!

So, what is the biggest time-bomb for Washington, for America, for capitalism, for the entire planet? No, it really not global warming ... not peak oil ... not even inequality. So, what's the absolute biggest, the one that's like the trigger mechanism on a nuclear bomb, the one that'll throw a monkey wrench into global economic growth, ending capitalism, even destroying modern civilization?

The one that, if not solved soon, renders *all* other efforts to solve *all* the other problems in the world, futile, irrelevant and virtually unsolvable?

News flash: Bill Gates "Billionaires Club" knows which one. Gates called billionaire philanthropists to a super-secret meeting in Manhattan back in 2009. Included: Buffett, Rockefeller, Soros, Bloomberg, Turner, Oprah and others meeting at the "home of Sir Paul Nurse, British Nobel Biochemist and President of the private Rockefeller University, in Manhattan," reported John Harlow in the *London Times Online.* During an afternoon session each was "given 15 minutes to present their favorite cause. Over dinner they discussed how they might settle on an 'umbrella cause' that could harness their interests."

So, what's the world's biggest time-bomb? Gate's billionaires unanimously agreed: "Population!" Don't solve Planet-Earth's over-population problem, and nothing else matters. Since their 2009 meeting, the world has added another half billion people, steadily marching to a predicted, unsustainable ten billion in 2050, one generation.

DIAMOND'S ALGORITHM FOR UNSUSTAINABLE PLANET-EARTH

"One of the disturbing facts of history is that so many civilizations collapse," warns Jared Diamond, an environmental biologist, Pulitzer prize winner and author of the classic *Collapse: How Societies Choose to Fail or Succeed.* Many "civilizations share a sharp curve of decline. Indeed, a society's demise may begin only a decade or two after it reaches its peak population, wealth and power."

Diamond's 12-part algorithm is quite simple, fits perfectly with the Pentagon's global warfare scenario: "More people require more food, space, water, energy, and other resources ... There is a long built-in momentum to human population growth called the 'demographic bulge' with a disproportionate number of children and young reproductive-age people." And if the 'bulge' stops for any reason, game over. Economic growth ends, killing capitalism.

So, look very closely: Diamond's algorithm has 12 coded time-bombs. But note, the first two are the primary triggers. The other 10 are derivative variables that are energized by these two primary triggers:

1. POPULATION GROWTH MULTIPLIER—UNSUSTAINABLE INCREASES

According to *Times Online:* A few months before the billionaires meeting Gates noted: "Official UN projections say the world's population will peak at 9.3 billion, but with charitable initiatives, such as better reproductive health care, we think we can cap that at 8.3 billion."

Can it be stopped? In a recent special issue of *Scientific American,* population was called "the most overlooked and essential strategy for achieving long-term balance with the environment." Why? Population's the new third-rail for politicians. So, they ignore it.

Yet, if all nations consumed resources at the same rate as America, we'd need six Earths to survive. Unfortunately, that scenario is unstoppable. Because by 2050, while America's population grows from 300 million to a mere 400 million, the rest of the world will explode from 6.3 billion to 8.9 billion, with over 1.4 billion each in China and India.

2. POPULATION IMPACT MULTIPLIER— CONSUMPTION INCREASING

Diamond warns: "There are optimists who argue that the world could support double its human population." But he adds, they "consider only the increase in human numbers and not average increase in per-capita impact. But I have not heard anyone who seriously argues that the world could support 12 times its current impact." And yet, that's exactly what happens with "all third-world inhabitants adopting first-world standards."

"What really counts," says Diamond, "is not the number of people alone, but their impact on the environment," the "per-capita impact." First-world citizens "consume 32 times more resources such as fossil fuels, and put out 32 times more waste, than do the inhabitants of the Third World." So, the race is on: "Low impact people are becoming high-impact people" aspiring "to first-world living standards." Yes, the American dream is now the global dream.

Get it? Our capitalist economy oversold The American Dream while creating new global markets and customers for our exports. Now everyone's got "The Dream!" Not just 310 million Americans, but 7.3 billion people worldwide are demanding more, more, more ... the new China Dream ... the India Dream ... the Africa Dream ... yes, "dreaming" has been America's biggest export for decades.

Warning: This "population impact multiplier" will also accelerate Diamond's global growth algorithm even with no population increases! Why? In Diamond's *Collapse,* the two key variables are what are called the Over-Population Multiplier and Population Impact Multiplier. Now let's closely examine how those two key variables impact all the other ten variables that are driving this algorithm:

3. FOOD PRODUCTION

Two billion people, mostly poor, depend on fish and other wild foods for protein. They "have collapsed or are in steep decline" forcing use of more costly animal proteins. The U.N. calls the global food crisis a "silent tsunami." Food prices rise making it worse for the 2.7 billion living below poverty levels on two dollars a day.

In "The End of Plenty," *National Geographic* warns that even a new "green revolution" of "synthetic fertilizers, pesticides, and irrigation, supercharged by genetically engineered seeds" may fail. Why? A joint World Bank/U.N. study "concluded that the immense production increases brought about by science and technology the past 30 years have failed to improve food access for many of the world's poor." And a *Time* cover story warns America's "addiction to meat" has led to farming technology that's "destructive of the soil, the environment and us."

4. WATER RESOURCES

Diamond warns: "Most of the world's fresh water in rivers and lakes is already being used for irrigation, domestic and industrial water," transportation, fisheries and recreation. Water problems destroyed many earlier civilizations: 'Today over a million people lack access to reliable safe drinking water." British International Development Minister recently warned that two-thirds of the world will live in water-stressed countries by 2015.

Water will trade like oil futures as wars are fought over water and other basic essentials noted earlier in Fortune's analysis of the Pentagon report predicting that warfare will define human life in this scenario of the near future.

5. FARMLANDS

Crop soils are "being carried away by water and wind erosion at rates between 10 to 40 times the rates of soil formation," much higher in forests where the soil-erosion rate is "between 500 and 10,000 times" replacement rate. This is increasing in today's new age of the 100,000-acre mega-fires and once-in-1,000-year disasters, dustbowls and droughts occurring every 100 days.

6. FORESTS

We are destroying natural habitats and rain forests at an accelerating rate. Half the world's original forests have been converted to urban developments. A quarter of what remains will be converted in the next 50 years.

7. CHEMICALS

All too often our solutions create more problems than they solve. For example, industries "manufacture or release into the air, soil, oceans, lakes, and rivers many toxic chemicals" that break down slowly or not at all. Consider the deadly impact of insecticides, pesticides, herbicides, detergents, plastics ... the list is endless.

8. ENERGY: FOSSIL FUELS & ALTERNATIVES

Pimco manages $747 billion for investors: equity, bonds and commodity funds. Manager Bill Gross once described a "significant break" in the world's "growth pattern." He's already betting we're past the "peak oil" tipping point. Consumer shopping will continue declining as economies grow very slowly in the future and "corporate profits will be static." A recent issue of *Foreign Policy* journal warns of the "7 Myths About Alternative Energy." Are biofuels, solar and nuclear the "major ticket?" Unfortunately, never greater that twenty percent say experts like David Yergin, a leading international energy consultant.

9. SOLAR POWER

Sunlight is not unlimited. Diamond: We're already using "half of the Earth's photosynthetic capacity" and we will reach the max by mid-century. In "Plundering the Amazon," *Bloomberg Markets* magazine warned that Alcoa, Cargill and other companies "have bypassed laws designed to prevent destruction of the world's largest rain forest ... robbing the earth of its best shield against global warming." Yes, free market capitalism may ultimately be the enemy of our survival.

10. OZONE LAYER

"Human activities produce gases that escape into the atmosphere" where they can destroy the protective ozone or absorb and reduce solar energy.

11. SPECIES DIVERSITY

"A significant fraction of wild species, populations and genetic diversity has been lost, and at present rates, a large percent of the rest will disappear in half century."

12. ALIEN SPECIES

Transferring species to lands where they're not native can have unintended and catastrophic effects, "preying on, parasitizing, infecting or outcompeting" native animals and plants that lack evolutionary resistance.

In spite of the clear message in Diamond's 12 time-bombs, he remains a "cautious optimist." What fuels his hope? Our leaders need "the courage to practice long-term thinking, and to make bold, courageous, anticipatory decisions at a time when problems have become perceptible but before they reach crisis proportions."

Unfortunately, Diamond says, history tells us that leaders are cautious and short-sighted, driven more by self-interest and nationalism than courage and long-term thinking. As a result, they get caught off guard and their worlds collapse, fast. They only respond to crises, usually too late.

9.29.09

CATASTROPHIC

ENDING

*"One of the most serious consequences of our actions is global warming.
Brought about by rising levels of carbon dioxide from the burning of fossil
fuels. The danger is that temperature increase might become self-sustaining
if it has not done so already ... We don't know where global warming will stop.
But the worst case scenario is that Earth will become like its sister planet
Venus with a temperature of 250 degrees Celsius and raining sulfuric acid.
The human race could not survive in those conditions."
Stephen Hawking, A Brief History of Time; The Grand Design*

STEPHEN HAWKING PREDICTS A "CATASTROPHIC ENDING"

Yes, Stephen Hawking predicts Planet-Earth's destiny—a "catastrophic ending" in the "not-too-distance future," warning us that the "future of the human race must be in space."

The good news, he wants you to become a hero, go save the world. Maybe work with the World Bank on developing nations, United Nations on climate change, maybe NASA, even the Peace Corps. Except they're all chasing a no-win scenario. An impossible dream with no viable solution. Why? Relentless opposition.

So go work for the "enemy?" For organization blamed for causing global warming. Yes, go change their culture *from the inside:* Big Oil, Big Ag, the GOP, fossil fuel billionaires, far-right conservatives, and all their climate-science-denying donors and lobbyists. Seriously. Maybe you'll be lucky, a new savior changing the deniers from within. Save Planet Earth ... before it's too late.

The clock is ticking faster, louder. Still all science deniers rally around a rock-hard faith in their capitalist ideology, change from within is probably a dead-end. Okay, so working for leftist idealists isn't working. And the hard-right's fighting too hard. You're in a warzone. Both headed for the same catastrophic ending Hawking predicts for Earth in the "not-too-distance future." What's the solution?

POPULATION GROWTH: JET FUEL FOR GLOBAL WARMING!

Listen to the World Bank scenario. Frustration is already replacing hope for one key official, Mark Cackler, the World Bank's manager for global agricultural and food security. You can feel his pain, like watching the Texas dust-bowl scenes on a dying Planet-Earth in the movie *Interstellar.* NASA was sending space explorers into several wormholes searching for new habitable planets to populate with future generations of Earth human species. Crackler's frustration with our planet's similarly impossible mission is apparent in reading recent op-ed piece in London's *The Guardian*:

The human race is "trapped in a vicious circle" according to the World Bank's agriculture manager, "we will need to grow 50% more food by 2050 to feed 9 billion people ... but agriculture, which is paradoxically vulnerable to climate change, generates 25% of heat-trapping greenhouse gas emissions that lead to climate change."

This is essentially a no-win scenario, a Zen Koan, the ultimate question that has no real answer: The more food "we grow using conventional methods, the more we exacerbate the problem. It's time for a climate-smart agriculture. But first we must address a few manmade problems." So here are the World Bank's four obstacles that must be solved before we can save Planet Earth from the "catastrophic ending:"

1. CLIMATE CHANGE AGREEMENTS FAIL WITHOUT AGRICULTURE POLICY

"There is a frustrating lack of attention paid to agriculture in the current global climate talks leading up to the Paris conference later this year," warns the World Bank's ag-manager:

"We need a climate change-agreement where agriculture is a big part of the solution, and delivers a triple win: higher agricultural productivity to feed more people and raise the incomes of poor farmers, especially women, greater climate resilience, and reduced greenhouse-gas emissions." But unfortunately, these goals are aggressively opposed by a powerful global network of capitalist corporations.

2. TECHNOLOGY IS FAILING AGRICULTURE AND FOOD PRODUCTION

"We still grow our food largely using 20th century technology, over 100 years old in the case of fertilizer production! We need more and better agricultural research to bring farming into the 21st century." But Silicon Valley isn't helping much. *MIT Technology Review* says America "Can't Solve Big Problems" anymore. Few tech geniuses want to be farmers. Elon Musk wants fast cars and powerful batteries. Apple wants more gadgets for our narcissistic consumer culture.

Mark Zuckerberg tried to go beyond, predicting five new technologies in the future of 10 billion "customers" on Planet Earth — virtual reality, artificial intelligence, telepathy, laser beams, even "Immortality," in answer to a question physicist Stephen Hawking called one of "the biggest questions in science." But Zuckerberg made no mention of agriculture, food or farming technology in the future, just more stuff for the global consumer-obsessed culture with short-term memories focused on selfies and the next-new-tech-thing.

Worse, Zuckerberg totally missed Hawking's biggest problem, his prediction of a "catastrophic ending" for Planet Earth in the not-too-distance future. "The long-term future of the human race must be in space," says Hawking, the kind of dark prediction that launched blockbuster films like *Avatar*, *Elysium*, and *Interstellar* which opened in a new farm belt dust bowl on a dying Planet-Earth.

Hawking warns that the human race better start planning now for living in space: "It will be difficult enough to avoid disaster on Planet Earth in the next hundred years, let alone the next thousand" and certainly not before America's space explorers and capitalist innovators, such as NASA and Elon Musk's SpaceX, find a home for human civilization on new planets and space travel there.

3. CARBON EMISSIONS KILLING US: STILL NO PRICING, TAXES, REGS

"Agriculture, like other sectors, wastes carbon because we don't price it properly," warns the World Bank's agriculture boss. "Carbon pricing is an essential way to cut greenhouse-gas emissions and lower climate risks." Yes, a "price on carbon can drive investment toward a greener agriculture, a cleaner economy and ultimately, more food for all."

Yes, a perfect solution, idealistic and hopeful, but it needs an unlikely surrender by hard-nosed Big Oil executives, shareholders, bankers and world's billion car owners pulling up to the gas pump every day.

Why? A *BusinessWeek* article detailed how Big Oil companies are already pricing in carbon taxes and regulation, if governments make the first move. But we also can predict — based on the experience of regulators with Wall Street banks — that Big Oil will do everything possible to oppose, fight, sabotage, kill or minimize any and all proposed carbon taxes and regulations.

4. WATER: AGRICULTURAL USE TRIGGERS 25% OF CO2 EMISSIONS

The World Bank's scenario is dark: Today agriculture "consumes 70% of the world's fresh water, and too much of it is wasted." We need to "become wiser about water." Because the "agricultural water generates 25% of heat-trapping greenhouse gas emissions that lead to climate change."

Ultimately our goal is to feed not just the seven billion living on Planet Earth today, but the 10 billion predicted by 2050. Yet, even the World Bank's agriculture manager admits that won't be easy: "Hunger has many causes, including ignorance and injustice and violence, and there is no single solution that will guarantee that every person, every day, everywhere, has enough to eat. But, as incredulous as it sounds, even though one in nine people go hungry today, it is within our power to eliminate extreme poverty and hunger by 2030."

Yes, like so many other idealists, hope springs eternal. And many are hopeful now, especially backed by Pope Francis' support of the UN climate-change initiatives. Moreover, the World Bank's agricultural manager is sees promise: "If the sustainable development goals and COP21 in Paris are opportunities to come up with bold ideas to tackle poverty, reduce inequality and address climate change, then climate smart agriculture should be acknowledged as one of those ideas that will enable us to do all three in one."

RIGID IDEOLOGIES VS ENDLESS ECONOMIC GROWTH

But will we get into space ... in time? How many? Mars colonies? The premise of the "Interstellar" film emerges from Hawking's prediction of a "catastrophic ending" for Planet Earth in the "not-too-distance future," and a clear warning that the "future of the human race must be in space."

Until then, we must feed Earth's exploding population at 10 billion by 2050 ... we do need innovative new 21st century agricultural research and technology ... and need more enlightened politicians and government leaders ... to give us "a global climate-change agreement where agriculture is a big part of the solution."

And finally, we need all the fossil fuel climate-science deniers to wake up and stop fighting change ... because their resistance is not only sabotaging the agricultural, farming and the food industry sectors. They're also accelerating the "catastrophic ending" Stephen Hawking predicts for Planet Earth in the "not-too-distant future."

But by then ... it'll be too late and innovative agricultural technology won't matter.

7.16.15

DENIALISM

FEARofDEATH

'"Sometimes facing up to the truth is just too hard. When the facts are distressing it is easier to reframe or ignore them. Around the world only a few have truly faced up to the facts about global warming ... It's the same with our own deaths; we all 'accept' that we will die, but it is only when our death is imminent that we confront the true meaning of our mortality." Clive Hamilton, Requiem for a Species: Why We Resist the Truth About Climate Change

'INVISIBLE HAND' OF CAPITALISM, SABOTAGING THE DREAM

"Working at Morgan Stanley is like making love to a gorilla," joked our president during his speech at an annual dinner years ago. "You don't stop when you want to. You stop when she wants to." Wives sat stoically. Our heads nodded, feigning laughter, not because it was funny, but because it was painfully true. And not just for us at Morgan Stanley, everywhere on Wall Street.

Yes, Wall Street is a savage jungle, gorillas rule, and every Wall Street investment banker at Morgan, Goldman, Citibank and the rest are all locked in this same lovemaking with that same Insatiable Gorilla.

You can't stop. You're trapped. So, you play. You feed on it. Blinded by the passion, the money, the power, the energy, the sense of life purpose she creates for you. Yes, she rules America's great capitalist jungle. You damn better obey.

Since those great days at Morgan Stanley I've come to a bizarre awareness: That "Insatiable Gorilla" is actually a metaphor for something profound. That Gorilla is the all-almighty "Invisible Hand" of capitalism Adam Smith immortalized in his classics on *The Wealth of Nations* and *The Theory of Moral Sentiments.*

Yes, the Insatiable Gorilla we all know is the mysteriously cryptic Invisible Hand guiding capitalism in the new century: the America that embodies Ronald Reagan's global superpower status, the democracy bred into me as a US Marine Corps sergeant who wanted to save the world from communism, and our economy and government that's balancing free-market conservative principles with liberal compassion, without self-destructing.

'INVISIBLE HAND' HIDES HUMAN'S DEEP FEAR OF DEATH

When I was new on Wall Street I read a bizarre assortment of books that enlightened me on the "Gorilla's 7 Laws." They included the 20th century new "Adam Smith's" *Super-Money, The Money Game* and his masterpiece *Powers of Mind.* Also, Napoleon Hill's *Success Through a Positive Mental Attitude,* Joseph Campbell's *Hero of a Thousand Faces,* Richard Bach's *Illusions,* Gail Sheehy's *Passages,* Alan Watts's *Way of Zen*, Scott Peck's *Road Less Traveled* and Robert Pirsig's Zen and the *Art of Motorcycle Maintenance.*

In fact, looking back all that was the only way to grasp the meaning of the delightfully enigmatic Invisible Hand. But the one book that's haunted me since my first days at Morgan Stanley was Ernest Becker's philosophical *Denial of Death.* It took several readings over the years to sink in, a little couch time while at Morgan Stanley, a doctorate in psychology and later working as a health-care professional helping a few hundred executives, physicians, actors, rock stars, athletes, politicians, royalty and other celebrity alums of the Betty Ford Center.

Today Becker's core message seems all too obvious and is indeed quite easy to understand for everyone except those trapped inside the bubble that has become Wall Street's great American Capitalist Jungle—all those who live in a jungle run by the Insatiable Gorilla that's a clever disguise for capitalism's Invisible Hand.

WALL STREET 'GORILLA' EXPOSES 'INVISIBLE HAND'

Stick with me, we're going to have an interesting time mixing, matching and merging metaphors so that we can penetrate Wall Street's behavior with the 'Insatiable Gorilla" as a channel to reveal the seven psychological principles of the banker's brain, quoting the wisdom of Earnest Becker and Sam Keen's brilliant preface summarizing Becker's message.

So, here's the short version of the "Gorilla's 7 Laws." Seven simple laws explaining the cycle of rising to power, self-sabotaging behavior at the peak, and inevitably self-destruction and collapse. Imagine we have an archetypal Wall Street insider, a banker here on the analyst's couch discovering for the very first time their deepest fears they hide, why humans make the decisions we make:

FIRST. AT BIRTH WE REALIZE THE WORLD IS HOSTILE, UNSAFE

Wall Street CEO, broker, trader with billions, they all face the same deep angst in their souls where an inner war rages, every day since birth. Becker's world is nothing like "Disneyland" says Keen: "Mother Nature is a brutal bitch, red in tooth and claw, who destroys what she creates," brutally "tearing others apart with teeth of all types — biting, grinding flesh," and more.

SECOND. FEAR OF DEATH OVERWHELMS US WITH INTENSE ANXIETY

We live in "terror: The harsh reality that "out there" are mortal enemies, out to destroy us. Becker sees our basic human motivation as a "biological need to control our basic anxiety, to deny the terror of death." So, we adapt, endure the pain of existence in a cruel world, endlessly racked with anxiety, "helpless, abandoned in a world where we are fated to die."

THIRD. WE CREATE CLEVER WAYS TO DENY FEAR, HIDE ANXIETIES

Yes, you must deny it absolutely, blocking the fears from your conscious awareness. To survive, to be productive, raise a family, you deny the harsh reality of your eventual death. Your brain is clever, is the "first line of defense that protects us from the painful awareness of our helplessness." So "we hide in our phony defense mechanisms" where "we feel safe … able to pretend that the world is manageable."

Unfortunately, "the price we pay is high," quoting Keen. "We repress our bodies to purchase a soul that time cannot destroy; we sacrifice pleasure to buy immortality … And life escapes us while we huddle within the defended fortress" of our false self.

FOURTH. 'TRANSCEND' DEATH AS AN IMMORTAL SAVING WORLD

Here's where the human mind is at its most brilliant: "Society provides the second line of defense against our natural impotence," says Keen. Yes, all cultures create "a hero system that allows us to believe we transcend death by participating in something of lasting worth. We achieve ersatz immortality by sacrificing ourselves to conquer an empire, to build a temple, to write a book, to establish a family, to accumulate a fortune, to further progress and prosperity, to create an information society and global free market."

In this rarified state of mind, we can even see that "corporations and nations may be driven by unconscious motives that have little to do with their stated goals." And driven by leaders whose unconscious motives have more to do with overcoming anxieties about death by proving they are heroes.

Where "making a killing in business or on the battlefield frequently has *less to do with economic need or political reality than with the need for assuring ourselves that we have achieved something of lasting worth.*" Yes, our leaders ostensibly pursue corporate, political and even altruistic goals while deep inside they're all selfish, self-seeking and narcissistic.

FIFTH. YOUR HEROIC JOURNEY AROUSES ENEMIES, BACKFIRES

Keen aptly summarizes this for the Wise Gorilla: "our heroic projects that are aimed at destroying evil have the paradoxical effect of bringing more evil into the world." America against China, GOP vs. Dems, etc.

But the real war is within us. Unfortunately, we are projecting it onto the people and the world around us. We fight harder to distract us from our fear of death, and convince ourselves we are indeed immortal, becoming as gods in our own mind. And yet, hiding deep under every hero quests to save the world is the old terror. And eventually, paradoxically, it backfires, opposition grows … new enemies return fire.

SIXTH. JOURNEY CREATES NARCISSISTIC HEROES & NEW ENEMIES

The great Gorilla of the Invisible Hand now focuses on Becker's opening paragraphs where we discover that "one of the key concepts for understanding man's urge to heroism is the idea of narcissism." In fact, it was Freud who "discovered that each of us repeats the tragedy of the mythical Greek Narcissus." We are "hopelessly absorbed with ourselves" and "twenty-five hundred years of history have not changed man's basic narcissism; most of the time … practically everyone is expendable except ourselves."

Ultimately, we realize we are driven not just to survive, but to strive to become as immortals, to merge with the gods. Our psyche, our brains, our very DNA has been programmed this way. And Wall Street bankers are uniquely self-centered narcissists. That's why capitalism is always "good," always will be, and is pre-wired in us.

SEVENTH. ENLIGHTENED HEROES? OR "NO EXIT" FROM JUNGLE

You want hope, asks the Insatiable Gorilla of the Invisible Hand? A way out of Wall Street's Jungle? An end to your bizarre Hero's Journey? You want solutions? A happy ending, new ways to get rich and retire in peace? Me too. But today, as with Sartre the existentialist, Becker offers "no exit" from your self-imposed hell. That's a great Zen paradox: You may achieve enlightenment, but even then, you're stuck here on Earth as a human, much as a Samurai warrior you must keep on fighting in the capitalists' jungle.

Unfortunately, that leaves us trapped in the inevitable conclusion that Wall Street's jungle of heroic narcissists will inevitably implode, sabotaging the Great America Dream. Yes, and same goes for all the world's private-equity billionaires, Silicon Valley philanthropists, all our too-big-to-fail bank CEOs, all our big-ticket wealth managers and brokers, multimillionaire high-frequency quants, media anchors, all the Super Rich, their lobbyists and their ideologically linked-at-the-hip politicians — all the narcissists in America's grand cultural conspiracy that's taken over Adam Smith's Invisible Hand, will inevitably sabotage America's capitalism.

Why? In the final analysis Wall Street insiders are like Icarus flying into the sun: Convinced they can avoid death by becoming as gods. But in so doing, their heroic journeys are destined to flame out, self-destruct, crash, burn, die. This is the fate of all individuals entrapped by the Gorilla of the Invisible Hand, a destiny we heard from many other visionaries. And obviously this virus has spread not just across Wall Street but all over America.

7.26.12

AMERICA'S

DEATHWISH

"Climate change isn't an 'issue' to add to the list of things to worry about, next to health care and taxes. It is a civilizational wake-up call, a powerful message—spoken in the language of fires, floods, droughts, and extinctions—telling us that we need an entirely new economic model and a new way of sharing this planet. Telling us that we need to evolve." Naomi Klein, This Changes Everything: Capitalism vs The Climate

20 RISKS AMERICANS FEAR DEADLIER THAN GLOBAL WARMING

A recent Gallup survey said only 36% of Americans think global warming will ever "pose a serious threat to their way of life in their lifetime." This year, the *Chapman University Survey of American Fears* began a new survey covering a list of 88 things that could trigger anxieties and fears, including "huge variety of topics ranging from crime, the government, disasters, personal anxieties, technology and many others."

Guess what? America's political parties are ideologically miles apart—with socialist Bernie Sanders warning that "climate change is America's biggest national security threat" and conservative politicians like Oklahoma Senator Jim Inhofe dismissing global warming as a "conspiracy" and "hoax"—still, it turns out that this new Chapman University "American Fears" poll is quite clear, only 30.7% of Main Street American respondents said global warming made them "afraid" or "very afraid," and that was before the recent elections.

Yes, the American public is aligned with the climate science-denying crowd. In fact, "global warming" didn't even make it into America's Top-20 Greatest Fears. Obviously, most Americans, most city-dwellers, most average folks in developed countries like

America appear to be tone deaf, tuning out comments in *Bloomberg News:* "New York Set to Reach Climate Point-of-No-Return in 2047."

Get it? Soon the overheating momentum of global warming will be so strong it will be irreversible, no matter how big the cutbacks, no matter what we legislate restraint, the planet will overheat indefinitely, eventually burning out like Mars. The *Washington Post* summarized the trend, with predictions of "temperatures too hot for Persian Gulf may be too hot for human survival," in less than two generations.

Still, Americans are dreamers, consistently tuning out or toning down climate threats. So here are some risky things Americans found far more dangerous than global warming. The scary fact is that while any one of these could trigger a major disaster in the short-term, global warming is virtually certain to trigger multiple global catastrophes in the long-run:

1. GOVERNMENT CORRUPTION (58%)
Money rules politics: Lobbying, elections, campaign fraud, payoffs, etc.

2. CYBER-TERRORISM, CYBER-WARS (44.8%)
China, Iran, North Korea, security experts say our energy grid will crash.

3. CORPORATIONS TRACKING PERSONAL DATA (44.6%)
Okay they need the big data in massive clouds but it's too easy to get mine.

4. TERRORIST ATTACKS (44.4%)
If this survey was taken after the Paris attacks, this would be number one.

5. GOVERNMENT TRACKING PERSONAL DATA (41.6%)
Expect lots more tracking to balance efforts to track potential terrorists.

6. BIOLOGICAL-CHEMICAL WARFARE (40.9%)
World saw Assad's brutal Syrian regime gas his own people. Next?

7. IDENTITY THEFT CRIMES (39.6%)
Hackers winning, easily grab passwords, job, tax, financial records.

8. ECONOMIC COLLAPSE (39.2%)
In 2000, Wall Street lost $8 trillion. In 2008, $10 trillion. Next after 2016.

9. RUNNING OUT OF MONEY IN RETIREMENT (37.4%)
Unfortunately, too many have too little savings, don't trust Social Security.

10. CREDIT CARD FRAUD (36.9%)
Credit cards added a new security chip. Counterfeiters upped their game.

So, there you go, the top fears worrying Americans in 2015. And where in global warming, climate change and the environment? Sorry activists, you did not even break into the top-20 ... American's have a secret death wish.

21. GLOBAL WARMING (30.7%)
Unfortunately, this ranking suggests rather weak public support for the
UN Climate Change Paris Accord signed back in late 2015.

Worse is the apparent subtle hypocrisy of the 197 global leaders signing the Paris Accord. A headline in the *Climate News Network* warned: "Biggest Economies still backing Fossil Fuels." Not only is the United States still "subsidizing" ExxonMobil and our other Big Oil companies to the tune of $4 billion a year, the "World's 20 leading economies give nearly four times as much in subsidies to fossil fuel production as total global subsidies to renewable energy."

And in the eyes of some key experts, this flaw could doom the future of the UN Climate Accord, as well as the future of our planet. *The Guardian* reported that James Hansen, the former head of NASA Goddard institute of Space Studies from 1981 to 2013 who made the Defense Department aware of the global warming threat, said of the Paris Accord signing:

"It's a fraud really, a fake ... just bullshit for them to say: We'll have a 2C warming target 'and then try to do a little better every five years.' It's just worthless words. There is no action, just promises. As long as fossil fuels appear to be the cheapest fuels out there, they will continue to be burned."

THE TURNING POINT

Still, we can hope. Maybe by the time America's 2020 presidential elections roll around, and current global warming trends continue accelerating the risks, disasters, agonies and the escalating costs, Republicans, Big Oil, the entire fossil fuel industry, and every climate science denier will be shocked awake, will surrender, stop fighting reality and get serious about the threat facing America and our entire way of life on this planet.

A year after the Chapman study, *Esquire* magazine reviewed a new Pew Research Poll concluding that conservatives believe "climate change isn't happening ... but even if it was, it wouldn't cause any really bad effects to the environment ... and even if it did, we couldn't do anything to help." As a result, conservatives generally do not support any policies or actions dealing with climate change and global warming. Neither do most humans, we just don't care.
10.25.15

QUESTION #2:

"CAN NEW TECHNOLOGY 'SAVE' OUR WORLD ... IN TIME?"

GENE POOL

MADNESS

"It might be the end of the world…by burning through coal and oil deposits, humans are putting carbon back into the air that has been sequestered for ten, in most cases hundreds of millions of years. In the process, we are running geologic history not only in reverse but at warp speed."
Elizabeth Kolbert, Sixth Species Extinction

SURVIVAL THE FITTEST: 'SELFISH GENE' BREEDS CAPITALISM

Will our gene pool ultimately self-destruct our home planet? Yes, because we're all capitalists. All with the same very selfish gene, the "capitalist gene" imbedded in our DNA, driving all humans. Today over seven billion. Within a generation, by 2050, ten billon humans, all with that mutating capitalist gene, the survival-the-fittest madness gene screaming: "Me *first!* Climate later."

That's human nature. Basic psychology, neuro-science and evolutionary biology. Survival the fittest. When the chips are down, our instinctual, fight-or-flight gut reaction for self-preservation wins every time. You protect yourself, your survival, your very existence. Worse, as Yural Harari warns in Homo Deus: a Brief History of Tomorrow:

Yes, you may also want a sustainable planet for future generations. Know you need it for the future. Maybe you drive a hybrid. Recycle. Eat organic. You may even be on a crusade to save the planet and save civilization from and impending catastrophe. Save the environment from global-warming disasters: melting arctic glaciers, rain forests disappearing, urban smog, toxic pesticides, dying species, deserts killing farm lands, ozone burning, lost energy reserves, diseases, pandemics … yes, all the right stuff.

But first, you must survive. You. A capitalist must survive if capitalism is to survive. It's in our DNA. Everybody's got it! Yes, we are all capitalists. And as a result, paradoxically, soon we will pass a point of no return, with 10 billion on Planet-Earth, with everybody demanding more, upping their lifestyle, burning energy, exhausting scarce resources—all driven by our inner survival instinct, that powerful, me-first, capitalist gene.

Why? All warnings about climate change, global warming and environmental threats will never be as immediate and strong as our "daily bread" needs, our thirst and hunger pains, our kids crying for a meal, a drink today.

'CAPITALIST GENE' WHISPERS, ME FIRST, NOW, CLIMATE LATER

Our capitalist gene parallels what evolutionary biologist Richard Dawkins saw in his classic work, *The Selfish Gene,* the natural evolutionary process passing genes along. Personality traits passing from parent to child, to species, to future generations. This Capitalist Gene is in the DNA of all humans, the survival of the fittest, where "more is never enough!'

And yes, it was there long before Adam Smith's theories, before Maynard Keynes, before Alan Greenspan and Paul Ryan embraced Ayn Rand's extreme capitalism, before Jack Bogle warned of "mutant capitalism" in his classic, *The Battle for the Soul of Capitalism.* The "capitalist gene" is in all humans, has evolved in there since the dawn of civilization.

When making economic decisions ... when the capitalist brain chooses between saving the future and getting personally richer now ... between saving the environment or paying a new tax or losing a benefit or right, paying an extra fee ... then the Capitalist Gene kicks in, and for most humans we're biased toward short-term self-interest. Self-preservation today is our first priority ... and to survive you better be the fittest!

The capitalist gene is our instinct for survival and instant gratification. It's in our brains, our blood, our DNA structure, motivating our rational thinking process, while still hoping someone, somehow, somewhere will eventually solve all the world's problems we created, will heal the world, someday, in the future, for future generations. While we take care of ourselves first. That's capitalism.

Don't believe me? That's natural. You're a capitalist, not some hard-core ideological climate-science denier. Skepticism is inherent in the capitalist mind-set, our brains, the collective conscience of capitalism. Trust yourself.

10 TYPES OF CAPITALIST: ALL DRIVEN BY THE "SELFISH GENE"

Ready? So, take a close look at the following profiles identifying the classic human Capitalist Gene in billions of capitalists across the world. See how many profiles fit you, fit America, fit the world's 7.3 billion humans. How many are driven by our selfish capitalist gene? You guessed it, everyone.

Now ask yourself: Who's going to cut back. First? Voluntarily? Which nation, farmer, entrepreneur, logger in the Amazon? Who? Who will choose without being forced by some government, catastrophe, global war, pandemic, drought, mega-fire, hunger?

Scan through these selected profiles of all the characteristics of capitalists around the world today. Chances are you will not only identify with at least one of these profiles. You'll also sense why all humans in our world have been, are now, and—when the chips are down, and it costs you some money—you will always be a capitalist today and an environmentalist tomorrow. So, see how many fit your daily life:

AUTOMOBILE CAPITALISTS & BIG OIL—ETERNAL LOVE FEST!

Not only do all humans need transportation, the automobile is a human status symbol. We have a deep love affair with our Mustang, Jaguar, Honda, and want a better one next. There are more than 1 billion autos in the world, used by billions of drivers. America has 240 million. China 80 million. And Big Oil just keeps riding on auto demand. A trillion dollars in annual revenues. Do you expect voluntary cutbacks to reduce carbon pollution? Don't bet on it.

CONSUMER CAPITALISTS: MORE IS NEVER ENOUGH

Rich, middle-class, poor, we all have a capitalist gene. 310 million Americans buy food, electronics, pay local taxes, drive the economy, make sure their kids get an education. Consumers want more money, goods, progress, a better future, and the hottest new gadgets. The American Dream is built on the capitalist gene, it was imbedded deep in our collective conscience before the American Revolution, powered the Industrial Revolution.

RETIREE CAPITALISTS: SAVE OUR SOCIAL SECURITY

Capitalism is fiercely competitive. AARP lobbyists fight for the best tax deals for 72 million boomers. Older folks are the fastest growing segment of global population, in America, China, throughout the world. All want security, earned entitlements, retirement nest eggs.

SHAREHOLDER CAPITALISTS: LOVE SHORT-TERM PROFITS

Our legal and political system has evolved to protect shareholders, executives and other insiders giving them the right to ignore long-term public interests in climate change, guaranteeing that as a separate legal person, a corporation's only responsibility is a duty to its stockholders' short-term interests, focusing on closing stock prices, quarterly earnings, annual bonuses. Regulation or taxation of carbon emissions must be resisted to protect shareholder profits. And climate problems are often just PR problems.

MIDDLE-CLASS CAPITALISTS: LABOR'S PIECE-OF-ACTION

Wall Street, CEOs, shareholders and the Super Rich want to cut corporate taxes and workers benefits. While our capitalist economy favors the moneyed class, the inequality gap is widening, and like the Crash of 1929 will trigger a revolution and a new depression, with our working class demanding a broader share of capitalism's rewards.

GOVERNMENT: BUREAUCRATS, POLITICIANS, LOBBYISTS

Who really runs America? The 537 politicians elected to the White House, Senate and Congress? No. A bizarre network of 261,000 lobbyists, over 5,000 appointed bureaucrats, plus millions of civil servants, postal workers, state-government employees, teachers, police, firefighters, private contractors, military offices and enlisted, all driven by the Capitalist Gene, over 40 million with personal interests in a continuing government payroll.

WORLD GOVERNMENTS, SOVEREIGN & INDEPENDENT

In *Every Nation for Itself: What Happens When No One Leads the World,* foreign policy expert Ian Bremmer shows us how the Capitalist Gene drives sovereign nations into competition worldwide. The U.S. competes with China's hybrid mix of capitalism, communism, socialist planning, state-owned banks, stock exchanges. Plus, there's fierce competition for global resources, like capital-rich/food-poor nations buying and hoarding worldwide agricultural lands for future domestic demands, depriving poor nations, setting up rebellions, wars, revolutions.

PHILANTROPHISTS: SELF-INTEREST OR PUBLIC INTEREST

Some philanthropists like The Gates are actually encouraging capitalism among farmers in poor nations, where farming is the primary employment. New micro-capitalists. Family planning and contraceptives are also freeing African farmers to increase incomes, tapping into the universal Capitalist Gene spirit of all farmers.

SILICON VALLEY CAPITALISTS: ENTREPRENTURES

A couple years ago *MIT Technology Review* asked, "Why Can't We Solve Big Problems?" The article made a strong point that new generation of entrepreneurs and leading technology minds may be willing to solve the tough challenges of the 21st century, but most of the ventures is focused on smaller problems and getting on a fast-track to personal wealth.

HUMANS INBREEDING A SELF-DESTRUCTIVE 'CAPITALIST GENE'

Can humans escape the destiny programmed in the DNA that's driving all of us? Or is it locked in, irreversible? Are Dawkins *Selfish Gene,* our "Capitalist Gene" and the mysterious "Madness Gene" in Kolbert's *Sixth Species Extinction* all one and the same, immutable, driving human behavior to eventual self-destruction? We know that sooner or later every individual human will die, that species go extinct, so what about nations, civilizations, planets? Does our cellular behavior also reveal our home planet's destiny?

All the warnings about melting glaciers, rain forests vanishing, toxic urban smog, pesticides, dying species, farm lands becoming deserts, ozone burning, lost energy reserves, diseases, pandemics and so much more won't matter much to capitalists in denial. In the final analysis, some kind of crash, collapse or wake-up call seems necessary to knock some sense into our brains, even jolt our capitalist gene pool into an evolutionary leap into a new dimension. But when that happens it will be too late.

1.12.14

INEQUALITY

GUILLOTINE

"Just months before the storming of the Bastille in 1789, everything was peachy. The social order ran smooth. The aristocracy partied…the next day they were being dragged through the streets by their frilly collars like common thieves…Even in the seconds before their heads were about to roll away from their bodies underneath the blade of the guillotine, it still puzzled the opulent Paris elite how this could be happening." Adbusters Magazine

WORLD'S INEQUALITY GAP WORSE THAN 1929, EVEN 1789

Good news for the richest 10%. Bad news for everybody else. The *Wall Street Journal* reports "the top 3% of families saw their share of total income rise to 30.5% in 2013 from 27.7% in 2010, while the bottom 90% saw their share fall." Yes, folks, the inequality gap just keeps widening to where 62 billionaires now "own the same as half the world," according to the 2017 Oxfam-Davos report.

The rich just keep getting richer. The rest just keep getting shafted, regardless of the long-term consequences to the American economy. The rich don't care. And based on how brazen the new GOP budget is about widening the inequality gap, you can bet they'll just keep thumbing their noses at the rest of America, as they did even after Pope Francis's historic speech before the U.S. Congress.

The inequality gap is now at 1929 levels, in fact, the widest it has been in a couple centuries, which makes the GOP gamble with the future of America a real "assault on the middle class," says CNN, as the GOP keeps adding more and more benefits for the wealthy, while cutting incentives for the 90% who are actually building the economy of America's future.

A couple years ago a *Credit Suisse Global Wealth Report* gave us a snapshot of just where this explosive inequality bubble is headed, reminding us of something far worse than the 1929 Crash, rather of the 1790s when inequality suddenly triggered the French Revolution, and 17,000 lost their heads under the guillotine.

The Credit Suisse data reveals that just 1% own 46% of the world, while two-thirds of the world's people have less than $10,000. *Forbes* also reported that just 67 billionaires already own half of Planet Earth's assets. Credit Suisse predicts our world will have eleven trillionaires families in two generations, as the rich get richer and the gap widens.

INEQUALITY GAP: TICKING TIME-BOMB & ECONOMIC GUILLOTINE

Can this trend continue? Or will it trigger a revolutionary economic guillotine? Nobel economist Joseph Stiglitz, author of *The Price of Inequality*, is not as optimistic as Credit Suisse: "America likes to think of itself as a land of opportunity." But today the "numbers show that the American Dream is a myth … the gap's widening … the clear trend is one of concentration of income and wealth at the top, the hollowing out of the middle, and increasing poverty at the bottom."

History is warning us: Inequality is a recipe for disaster, rebellions, revolutions and wars. Not in two generations. Much, much sooner, a reminder of the Pentagon's famous 2003 prediction: "As the planet's carrying capacity shrinks, an ancient pattern of desperate, all-out wars over food, water, and energy supplies will emerge … warfare will define human life on the planet by 2020." Yes, much sooner than two generations.

Early warnings of a crash are dismissed over and over ("a temporary correction"). They gradually numb us about the big one. Time after time we forget history's lessons. Until finally a big surprise catches us totally off-guard. Financial historian Niall Ferguson put it this way: Before the crash, our world seems almost stationary, deceptively so, balanced, at a set point. So that when the crash finally hits, as inevitably it will, everyone seems surprised. And our brains keep telling us it's not time for a crash.

Till then, life just goes along quietly, hypnotizing us, making us vulnerable, till shockers like Bear Stearns or Lehman Brothers upset the balance. Then, says Ferguson, the crash is "accelerating suddenly, like a sports car … like a thief in the night." It hits, shocks us wide-awake. In our denial, we may keep telling ourselves it's just another short-term correction in a hot bull market. Until suddenly, it's accelerating, a Mack truck hits.

FRENCH REVOLUTION, HISTORY LESSON NEVER LEARNED

Angry masses, let resentment build, fuming inside. Their Treasury was bankrupt. High interest on national debt consumed half their tax revenues. Why? Earlier wars, a decedent aristocracy, an incompetent King Louis XVI. The anger so intense that during the 1792-93 "Reign of Terror" even the King was guillotined, along with 17,000, many who were innocent, as inequality ripped apart the France nation.

Why? The aristocracy, intellectuals and the rich were oblivious of the needs of the masses, much like our leaders today. As *Adbusters* magazine put it: "Even in the seconds before their heads were about to roll away from their bodies underneath the blade of the guillotine, it still puzzled the opulent Paris elite how this could be happening." Yes, they were clueless till the end, in denial, not listening to the masses for many years. Like today across America.

The truth is, revolutions catch nations by surprise: "Just months before the storming of the Bastille in 1789, everything was peachy. The social order ran smooth. The poor paid their dues. The middle class kept their mouths shut. The aristocracy partied ... next day they were being dragged through the streets by their frilly collars like common thieves."

14 TRIGGERS THAT CAN IGNITE NEW INEQUALITY FIREBOMB

How close are we to a new Bastille Day? Barry Ritholtz's *The Big Picture* website posted "The Stunning Truth About Inequality In America," a list of 14 triggers from "WashingtonsBlog," warning us the inequality gap is accelerating so rapidly, widening so fast that America may soon be at what we call Bastille Day levels, an inequality gap so great it is the fuel and trigger that can ignite an angry people into revolution.

These 14 triggers are reinforced by the statistics in all the reports from the latest Federal Reserve study, Credit Suisse Reports, the Pentagon's 2020 prediction, Pope Francis' "Apostolic Exhortation," and similar ones from Naomi Klein, Chris Hedge, Niall Ferguson and other climate change authorities. Here's an edited must-read summary of the "Stunning Truth About Inequality:"

1. Inequality gap's worse than you imagine. "Americans consistently underestimate the amount of inequality in our country ... would be shocked to learn the truth...
2. Gap's worse than history's worst. "Twice as bad as in ancient Rome, worse than in tsarist Russia, worse than in America's Gilded Age."
3. America is falling behind other developed nations. "Worse in America than any other developed nation." Could ignite a 1790's style revolution.
4. Inequality permanent. "Staggering inequality in America has become permanent."
5. America's two economies. "There are two economies: one for the rich, and the other for everyone else." Thomas Piketty's "Capital in the Twentieth Century," confirms it.
6. Top 1% in a bull rally, while the 99% in recession. "Economy's recovered for the richest 1% ... the rest of the country is more or less stuck in a depression."
7. Rich keep getting richer. "The super-rich are raking in more than ever before."
8. Poor getting poorer. "While more and more people are sliding into poverty."
9. Middle class now dead. "One of every five households in the America is on food stamps. The middle class has more or less been destroyed."
10. Inequality causes crashes. "Who's who of prominent economists and investors say that inequality causes crashes and hurts the economy." Start preparing now.
11. Great Depression. "Extreme inequality helped cause the Great Depression ... current financial crisis ... fall of the Roman Empire."
12. Bad politicians and policies. "Inequality isn't happening for mysterious or uncontrollable reasons. Bad government policy is responsible for runaway inequality."

13. Leadership. "Bush was horrible, but income inequality has increased even more under Obama than under Bush."
14. Conservatives. "It's a myth that conservatives accept runaway inequality. Conservatives are as concerned as liberals regarding the stunning collapse of upward mobility, the ending of the American Dream."

In *Wealth, War and Wisdom,* Barton Biggs, long-time Morgan Stanley global strategist warns of the "possibility of a breakdown of the civilized infrastructure." He advised his super-rich clients to get prepared for a post-apocalypse era: Buy a farm up in the mountains: "Your safe haven must be self-sufficient and capable of growing some kind of food ... well-stocked with seed, fertilizer, canned food, wine, medicine, clothes, etc. Think Swiss Family Robinson."

Astrophysicists warns that the "future of our civilization is in space." So ask yourself: Will there be any "safe havens" on Planet-Earth after the next Bastille Day revolution?

4.20.15

MUTANT

CAPITALISM

The "psychological man of our times, the final product of bourgeois individualism, which in its decadence has carried the logic of individualism to the extreme of a war of all against all, the pursuit of happiness to the dead end of a narcissistic preoccupation with the self…demands immediate gratification and lives in a state of restless, perpetually unsatisfied desire." Christopher Lasch, The Culture of Narcissism: American Life in an Age of Diminishing Expectations

HOW NARCISSISM'S KILLING ADAM SMITH'S MORAL CAPITALISM

Atlas Shrugged is Ayn Rand's most popular work. A mysterious rebel leader John Galt saves America from economic ruin. But before we rise from the ashes, before the redemption, comes the blowup. And for that, let's turn to my Ayn Rand favorite, *The Fountainhead,* where Rand offers a subtle hint of capitalism's rise and eventual demise and self-inflicted death.

In *The Fountainhead* Howard Roark is the ultimate individualist, an idealistic architect and archetypal free-market capitalist. Enraged when second-rate competitors compromise the integrity of his plans for a modern building, Roark seeks revenge, takes the law into his own hands and, in a terrorist act, sneaks onto the construction site in the dark of night and dynamites the building — *kaboom* — destroying it!

Flash forward: You have a perfect metaphor for Rand's extreme ideology, how today it is turning against America, blowing up capitalism itself, self-destructing as the excesses of a great ideology spin out of control, choking on the very dreams that fueled it for decades.

Yes, soon the commanding inner voice of Rand's super-capitalism that's imbedded itself deeply in America's collective unconscious will self-destruct capitalism, in a volley of excesses, taking down the market, triggering a total economic collapse and profoundly altering America's political destiny as the global superpower.

AYN RAND'S SOLUTION: DON'T LIKE SOMETHING? BLOW IT UP!

Capitalist ideologues are in a perpetual cultural war for the soul of America, fighting society's "moochers, looters and parasites," anyone and everyone who is demanding government money to solve their problems. The elite see America degrading into a socialist welfare state and communism. Rand says capitalism must be free:

> *"When I say 'capitalism,' I mean a pure, uncontrolled, unregulated laissez-faire capitalism, with a separation of economics, in the same way and for the same reasons as a separation of state and church." Why? Because "capitalism is the only system that can make freedom, individuality and the pursuit of values possible in practice because capitalism demands the best of every man, his rationality, and rewards him accordingly. It leaves every man free to choose the work he likes, to specialize in it, to trade his product for the products of others, and to go as far on the road of achievement as his ability and ambition will carry him."*

Paradoxically, Rand's capitalist ideals have gone far off the course defined by Adam Smith, warping into an unwritten conspiracy binding conservative politicians and the egocentric excesses of Wall Street, Corporate CEOs, and the Forbes 400 Super Rich.

WEALTH, RELIGION, LIBERALISM, ALTRUISM, MORALS

In a recent USA Today op-ed piece, "Ayn Rand and Jesus," Stephen Prothero, a Boston University professor of religion and author of *God is Not One: The Eight Rival Religions That Run the World—And Why Their Differences Matter,* poses a serious challenge to Rand's disciples: "Idolatry of the conservative icon should lead to some soul-searching within the GOP. After all, Christian morality has no place in an 'Atlas Shrugged' world."

Prothero's list of "Rand's adoring acolytes" includes Wisconsin Congressman Paul Ryan, Texas Rep. Ron Paul and Kentucky Sen. Rand Paul, all conservatives. But while "Ayn Rand is the GOP's new savior, no one seems to be taking notice of just how opposed their two philosophies are."

For Rand, the war is not between "God vs. Satan, but individualism vs. collectivism. While Jesus says, 'Blessed are the poor,' she sings hosannas to the rich. The heroes of 'Atlas Shrugged' …are the captains of industry … The villains are the looters and moochers," people who by hook (guilt) or by crook (government coercion) steal from the hard-won earnings of" rich capitalists.

Get it? Rand turns "traditional Christian morality" on its head: "Altruism is immoral, and selfishness is good. Moreover, there isn't a problem in the world that laissez-faire capitalism can't solve if left alone to perform its miracles." In Rand's quixotic world, capitalists are the new saviors, performing miracles.

But can you imagine Jesus ever saying that? No wonder Prothero says "Rand's work reads to me like a vulgar rationalization for greed lying on top of a perverse myth," leaving him surprised "at how few GOP thinkers seem to see how hostile her philosophy is to conservatism itself." Why indeed, because there's nothing Christian about Rand's defense of a soulless capitalism that's not only lacking in traditional Christian compassion but has become totally narcissistic.

ADAM SMITH'S MORALITY VS EGOCENTRIC NARCISSISM

The *Utne Reader* quoted from Christopher Lasch's groundbreaking "Culture of Narcissism," published a generation ago at the dawn of Reaganomics: The economic man "has given way to the psychological man of our times — the final product of bourgeois individualism, which in its decadence has carried the logic of individualism to the extreme of a war of all against all, the pursuit of happiness to the dead end of a narcissistic preoccupation with the self." There he "demands immediate gratification and lives in a state of restless, perpetually unsatisfied desire."

Prothero reminds us "real conservatism is also about sacrifice." Today however, the new narcissists ruling conservatism and capitalism "will brook no such sacrifice. Serve yourself, she tells us, and save yourself as well. There is no higher good than individual self-satisfaction."

And yet "one of the reasons we are in our current economic quagmire is that none of our leaders is willing to ask us to sacrifice. Democrats call for more spending and more taxes; Republicans call for lower taxes and less spending, and what we get is the most fiscally ruinous half of each: lower taxes and more spending."

Prothero's challenge: His "aim is to force a choice. If you are going to propose a Robin Hood budget, you have to decide whether you are robbing from the poor to give to the rich, or robbing from the rich to give to the poor. Because you cannot do both." Yes, America is "a free country. Just don't tell me you are both a card-carrying Objectivist and a Bible-believing Christian. Even Rand knew that just wasn't possible."

"THE BATTLE FOR THE SOUL OF CAPITALISM" WAS LOST

Future historians will see Ayn Rand as both the patron saint of the new narcissistic capitalism that has dominated conservatives and Reaganomics for the last generation. Her legacy will also include the "Death of Capitalism," a demise Jack Bogle wrote about so eloquently several years earlier in his *Battle for the Soul of Capitalism*.

Unfortunately, the new narcissistic capitalists are blind to this paradigm shift in America's destiny. As Lasch puts it: "Impending disaster has become an everyday concern, so commonplace and familiar that nobody any longer gives much thought to how disaster might be averted. People busy themselves instead with survival strategies, measures designed to prolong their own lives, or programs guaranteed to ensure good health and peace of mind." Get it? Deep inside we know it's too late, so we give up, double down, go deeper into materialism.

Yes, this narcissism is metastasizing so rapidly Americans feel ever more helpless to solve our problems, making collapse a self-fulfilling prophecy. Yes, this toxic narcissistic virus has infected America's soul, eating away at our core values while blinding us to both the problem and the solution.

We have lost the collective spirit that led 57 capitalists to risk their lives and fortunes by signing the Declaration of Independence. That's spirit's dead. Today it's "every man for himself" in a dark capitalist anarchy.

You ask, why do we embrace our own demise like out-of-control addicts? In behavioral economics, as in classical Greek drama, Jungian psychology and cultural mythologies ... all the battles we see "out there" are actually projections of unresolved conflicts raging deep within our own souls ... we're rehashing old traumas projected on the outside world as battles between our highest ideals and our darkest secrets ... classic battles between good and evil.

But they are conflicts buried deep in what Jung called "The Shadow," a prison of dark secrets we cannot admit even to ourselves. In there, fierce battles are fought for the possession of our immoral souls ... projected onto news, politics and finance, in television and films, theater, literature, history and dreams, at the dinner table and in the bedroom, "out there" we try to resolve our innermost secrets, never fully understanding how our minds are tricking us into inaction.

And as our individual souls and our collective unconscious mind splits further and further apart, eventually we will collectively implode and collapse.

6.14.11

G-ZERO

GLOBAL ANARCHY

"For the first time in seven decades, there is no single power or alliance of powers ready to take on the challenges of global leadership. A generation ago, the United States, Europe, and Japan were the world's powerhouses, the free-market democracies that propelled the global economy forward. But today, they struggle. This leadership vacuum is here to stay... call it geopolitical creative destruction or just the sound of things falling apart." Ian Bremmer, Every Nation for Itself: What Happens When No One Leads the World?

THE 'G-ZERO' ECONOMY: SIX STAGES TRIGGER GLOBAL ANARCHY

G-Zero is the "new world order in which no single country or durable alliance of countries can meet the challenges of global leadership. What happens when the G20 doesn't work and the G7 is history." asks geopolitical economist Ian Bremmer in his bestseller, *Every Nation for Itself: What Happens When No One Leads the World?*

The answer is simple: Capitalists are filling the vacuum, take leadership control, replacing presidents, prime ministers, kings. This new reality was highlighted in a photo-op with the $40 million a year Exxon Mobil president shaking hands with the CEO of Rosneft, Russia's oil producing giant. The deal: a $700 million oil exploration pact in the Arctic.

What irony: The U.S. president makes just $400,000 a year. He's under pressure from environmentalists about Big Oil. We're in a new Cold War with Russia, issuing weak sanctions. And here's one of America's biggest capitalists thumbing his nose at politics, smiling, shaking hands, getting a deal to drill oil with an archenemy in the threatened Arctic Ocean against the warnings of environmentalists worldwide.

1. LEADERS OF MORE NATIONS ADOPTING RULES OF CAPITALISM

Yes, economics trumps politics: Bremmer warns, in our 21st century world, no single nation is willing and capable of leading the rest. Result: Anarchy — *economic, political and moral anarchy* — with 197 nations all selfishly acting out of self-interest. And worse, there's no one to serve as the economic, political and moral compass for the rest of the world:

"If the worst threatened — a rogue nuclear state, a major health crisis, the collapse of the system — where would the world look for leadership?" asks Bremmer. No effective political leaders around: "For the first time in seven decades, there is no single power or alliance of powers ready to take on the challenges of global leadership. A generation ago, the United States, Europe, and Japan were the world's powerhouses, the free-market democracies that propelled the global economy forward. But today, they struggle just to find their footing."

Bremmer goes further, warning, "This leadership vacuum is here to stay, as power is regionalized instead of globalized. Now that so many challenges transcend borders, from the stability of the global economy and climate change to cyber attacks and terrorism, the need for international cooperation has never been greater," but missing in action.

Bremmer answered this pivotal question in earlier book, *The End of the Free Market: Who Wins the War Between States and Corporations?* "There is a second critical element defining our new "G-Zero World Order."

2. CHINA'S HYBRID SYSTEM—BIG STEP TO TOTAL CAPITALISM

Bremmer sees sovereign states like China in an earlier stage of transition, combining a little capitalism with their traditional government style: "A number of authoritarian governments, drawn to the economic power of capitalism but wary of uncontrolled free markets, have invented something new: state capitalism. In this system, governments use markets to create wealth that can be directed as political officials see fit."

So who wins? Imagine "G-Zero" as a chess match: anarchy is the queen in a virtual no-man's-land of regional battlefields with 197 nations fighting in multiple skirmishes, each serving their own special interests, with no international overseers. Moreover, in this game, anarchy has a powerful partner; capitalism is power, and more and more capitalists are filling the vacuum left by ineffective government leaders.

Bremmer has been observing the "rise of state capitalism and its long-term threat to the global economy. The main characters in this story are the men who rule China, Russia, and the Arab monarchies of the Persian Gulf, but their successes are attracting imitators across much of the developing world.

Wrong: "State Capitalism" is not the next big trend. Forget states ... forget governments ... forget nations ... forget democracy ... think economics ... think capitalism. Capitalism is nullifying and taking over the functions of governments, of nations, of the global economy.

3. MORE CAPITALISTS REPLACING POLITICAL LEADERS

Open your eyes: Capitalism is the one and only next big trend, has been for a generation. No, not the pure competitive 1776 capitalism Adam Smith envisioned in his classics *The Theory of Moral Sentiments* and *The Wealth of Nations.*

Today we have a new ultraconservative capitalism dominating, manipulating and controlling titular government leaders with the ideologies of Milton Freidman, Ayn Rand and Ronald Reagan, as they have been distorted, morphing into today's obstructionist GOP and its do-nothing tea party allies. But always lurking behind in the shadows, superrich capitalists pulling are the real strings of power.

Yes, the new capitalism has conquered the world in one short generation, fueled by many traits: anarchy, gridlock, individualism, egocentricity, arrogance and a blind faith in greed. And that spells future troubles for superrich capitalists and the rest of the world.

Here's how Bremmer and economist Nouriel Roubini explored this rapid acceleration of the last generation that's redirected capitalism into a new morally deficient anarchy. They previewed the new world order in *Foreign Affairs* a few years ago, warning us of the "G-Zero World: The New Economic Club Will Produce Conflict, Not Cooperation."

4. CAPITALIST COMPETITION LEADING INTO CONFLICTS

Living in our new G-Zero world where no one can set "a truly international agenda" will intensify "conflict on the international stage over vitally important issues, such as international macroeconomic coordination, financial regulatory reform, trade policy, and climate change. This new order has far-reaching implications for the global economy, as companies around the world sit on enormous stockpiles of cash, waiting for the current era of political and economic uncertainty to pass. Many of them can expect an extended wait."

Why? Conflicts, competition, wars and revolutions will keep accelerating in the new G-Zero world order, with no strong leaders among the 197 nations solving the big issues; by default, we will fulfill the Pentagon's 2020 prediction in *Fortune:* "As the planet's carrying capacity shrinks, an ancient pattern of desperate, all-out wars over food, water, and energy supplies would emerge ... warfare is defining human life."

So forget cooperation. Today globalization has created a massive arena for the new anarchism fueled by capitalist ideologues perpetually at war, fighting to gain the prize by defeating their opponents, riding the thrill of victory, avoiding the agony of defeat.

5. CAPITALIST LEADERS ACCELERATE TO MILITARY ACTION

So what's new? With the global power of both anarchy and capitalism expanding, our new leaderless G-Zero World Order is vulnerable to new risks. As Bremmer and Roubini put it: "There is nothing new about this bickering and inaction."

They remind us that world powers took four decades to negotiate the Nuclear Nonproliferation Treaty. "In fact, global defense policy has always been essentially a zero-sum game, as one country or bloc of countries works to maximize its defense capabilities in ways that deliberately or indirectly challenge the military preeminence of its rivals."

But today government leaders are impotent, rapidly being replaced by superrich, powerful capitalists in the new G-Zero global economy where the rules of the game are rapidly changing: "International commerce is a different game; trade can benefit all players."

Wrong. It can, once was, but no longer, not today: "In the past, the global economy has relied on a hegemon, the United Kingdom in the eighteenth and nineteenth centuries and the United States in the twentieth century, to create the security framework necessary for free markets, free trade, and capital mobility."

No hegemon in the G-Zero World Economy. We have a vacuum of international leadership just at the moment when it is most needed." And in the absence of powerful political leaders, capitalists are filling the leadership vacuum ... economic inequality is increasing ... setting the stage for more wars ... widespread rebellions ... and new class revolutions.

6. CAPITALISM + INEQUALITY = ANARCHY + REVOLUTIONS

The world mass-producing billionaires. Today there are over 1,800 billionaires (and rising) compared with 322 in 2000. All in one generation. *Forbes* says the U.S. has 492, Europe 468, China 358, Russia 111, Latin America 85, India 65, Canada 29, Africa 29, and more. Today 67 billionaires have as much wealth as the poorest 3.5 billion half of the world's population. And the richest of the Super Rich will get even richer: Credit Suisse Bank predicts eleven trillionaire families by 2100.

Finally, while billionaires and bankers see perpetual economic growth as a given, as population explodes from 7.3 billion to 10 billion, these capitalists more specifically see these billions as new consumers feeding economic growth, all reinforcing their belief in the trend and adding more billionaires. Unfortunately, sooner or later the leaders of this G-Zero Capitalist World will realize that perpetual economic growth is impossible on a planet of limited resources.
8.27.14

BIG PROBLEMS!

TWITTER BRAINS

*"It's not true that we can't solve big problems through technology,
we can. But all these elements must be present: political leaders and the
public must care to solve a problem, our institutions must support its
solution, it must really be a technological problem, and we must understand
it. The Apollo program, which has become a metaphor for technology's
capacity to solve big problems, met these criteria, but it is
an irreproducible model for the future. This is not 1961."
Jason Pontin, Editor-in-chief, MIT Technology Review*

5 REASONS TECHNOLOGY CAN'T SAVE OUR PLANET

"You Promised Me Mars Colonies," warned astronaut Buzz Aldrin on the cover of the *MIT Technology Review*. "Instead, I Got Facebook!" And there, folks, you have the anti-Silicon Valley argument in a nutshell, succinctly put by the *Review's* editor-in-chief Jason Pontin in his provocative article, "Why We Can't Solve Big Problems."

Why? One answer: Because America's leaders no longer think long-term, no longer invest in "Big Problems" like landing a man on the moon. So, Silicon Valley, and the rest of America's top innovative technology geniuses, focus on "little problems," short-term money-makers like Facebook and friends, Twitter, Linkedin, Snapchat and Instagram. Even Bezos and Musk's race to build Mars colonies for a few million humans is at best another distraction from focusing on the 'Big Problems" facing the billions left behind on our rapidly overcrowding planet.

Back in 1972 Pontin was just five as Apollo 17 lifted off: "My mother admonished me not to stare at the fiery exhaust of the Saturn 5 rocket. I vaguely knew that this was the last of the moon missions — but I was absolutely certain that there would be Mars colonies in my lifetime. What happened?"

Yes, what *did* happen? Why did America fail Buzz Aldrin? Fail Silicon Valley? Fail America's next generation? Fail to keep Kennedy's vision alive: Why is America failing as the world's technology leader? What happened to our vision?

AMERICA'S FUTURE: UNICORNS, APPS, MEGALOMANIA

"That something happened to humanity's capacity to solve big problems is a commonplace," says Pontin. "There is a paucity of real innovations. Instead, they worry, technologists have diverted us and enriched themselves with trivial toys."

Guys like PayPal co-founder Peter Thiel are echoing Aldrin: "We wanted flying cars — instead we got 140 characters." Pontin even says "Thiel is caustic," he doesn't even "consider the iPhone a technological breakthrough compared to the Apollo program." The Internet is "a net plus — but not a big one." And yes, Twitter is "job security for the next decade," for over 3,500 people. But Thiel, a billionaire and one of the best minds in the Valley, asks the big question: "What value does it create for the entire economy?"

Apparently not much to Silicon Valley critics: "We should be aiming higher." From the Founder's Fund manifesto, "What Happened to the Future?" They companies and toward companies that solved incremental problems or even fake problems." Instead of being a "funder of the future," Silicon Valley had "become a funder of features, widgets, irrelevances."

SILICON VALLEY CREATED "BIG PROBLEMS" ROADBLOCK

The criticisms are misleading, says Pontin: "The argument that venture capitalists lost their appetite for risky but potentially important technologies clarifies what's wrong with venture capital and tells us why half of all funds have provided flat or negative returns for the last decade."

But "even during the years when VCs were most risk-happy, they preferred investments that required little capital and offered an exit within eight to 10 years." Truth is, "VCs have never funded the development of technologies that are meant to solve big problems," says Pontin, in a direct challenge to Silicon Valley's so-called 'Big Problems' syndrome.

And that forces him back to the core question raised by Buzz Aldrin, Peter Thiel and every high-tech investor more interested in quarterly profits and big payoffs: "Putting aside the personal-computer revolution, if we once did big things but do so no longer, then what changed?" Pontin does a brilliant job diagnosing five key Big Problem macro-trends:

1. MORE IMPORTANT PUBLIC POLICY INVESTING ON EARTH

Seriously, is Silicon Valley's 'Big Problem' a real problem? Or did their high-tech geniuses make it up? Pontin says "sometimes we choose not to solve big technological problems. We could travel to Mars if we wished. NASA has the outline of a plan," and "if the agency received more money ... humans could walk warn that "venture investing shifted away from funding transformational on the Red Planet sometime in the 2030s." But "we won't, because there are, everyone feels, more useful things to do on Earth." Indeed, efforts by Elon Musk and Jeff Bezos can't outrun Earth's ticking population clock.

2. ECONOMIC COSTS HUGE CONSTRAINTS ON TECHNOLOGY

Energy is a real Big Problem with real cost issues: "Less than 2% of the world's energy consumption was derived from advanced renewable sources such as wind, solar and biofuels." Still, "coal and natural gas are cheaper than solar and wind ... and petroleum is cheaper than biofuels." Another: Yes, climate change is a real Big Problem. But because the main cause of global warming is carbon dioxide released as a by-product of burning fossil fuels, we need renewable energy technologies that can compete on price with coal, natural gas and petroleum. At the moment, they don't exist." Yet Bill Gates still presses for an "energy miracle" that seems as far away as the black holes in Avatar and Interstellar.

3. DOMESTIC, GLOBAL POLITICAL IDEOLOGIES BLOCK ACTION

Yes, "economists, technologists and business leaders agree on what national policies and international treaties would spur the development and broad use of such alternatives," says Pontin. But without agreement on how "to share the risks of development, alternative energy sources will continue to have little impact on energy use, given that any new technology will be more expensive at first than fossil fuels." Worse, there's "no hope of any U.S. energy policy or international treaties" given conservative opposition to regulations and because developing markets like China and India will not reduce their emissions without offsetting reductions by developed nations.

4. MOST "BIG PROBLEMS" ARE NEVER TECHNOLOGICAL

Most Big Problems aren't technology problems, "or could more plausibly be solved through other means." We used to believe famines were problems in food supply, etc. But Nobel laureate economist, Amartya Sen proved that "famines are political crises that catastrophically affect food distribution."

Pontin warns: The hope for technological solutions is "very seductive — so much so that disappointment with technology is inevitable." Malaria resists technology solutions. "The most efficient solutions are simple: eliminating standing water, draining swamps, providing mosquito nets and, most of all, increasing prosperity." But that didn't stop Bill Gates and other technologists from searching for "ingenious solutions ... But they all suffer from the vanity of trying to impose a technological solution on what is a problem of poverty."

5. BIG PROBLEMS TOO COMPLEX ... NOBODY UNDERSTANDS

Finally, warns Pontin, "sometimes big problems elude any solution because we don't really understand the problem." Get it? Us humans are the real "Big Problem!" Early in the 1970s biotechnology breakthroughs flowed. Later breakthroughs were tougher. We just don't "understand the fundamental biology of many diseases." In 1971 Nixon declared a war on cancer, "but we soon discovered there were many kinds of cancer" that resist treatment. Gene sequencing is now helping us understand the different cancers and "how their mutations express themselves in different patients."

Another big one: As nations age, the "dementia plague" is becoming "the world's most pressing health problem. By 2050, palliative care in the United States alone will cost $1 trillion a year. Yet we understand almost nothing about dementia and have no effective treatments." And we've been fighting a war on drugs for decades and it just gets worse.

SILICON VALLEY'S "BIG SOLUTIONS" TO BIG PROBLEMS?

In the end, Pontin is realistic: "It's not true that we can't solve big problems through technology; we can," he admits. "But all these elements must be present: political leaders and the public must care to solve a problem, our institutions must support its solution, it must really be a technological problem, and we must understand it. The Apollo program, which has become a metaphor for technology's capacity to solve big problems, met these criteria, but it is an irreproducible model for the future. This is not 1961."

Bottom line, Jason Pontin's full article, "Why We Can't Solve Big Problems" in *MIT Technology Review* may have been for tech insiders. But it is also a must-read for Wall Street and Main Street investors looking for hi-tech investments. He challenges and analyzes the 'Big Problems' raised by astronaut Buzz Aldrin and Silicon Valley's best and brightest minds. It's guaranteed to get you thinking about the role of technology in America's future as a superpower that will succeed in saving the world … from itself.
12.21.12

CREATIVE

DESTRUCTION

"There was virtually no growth before 1750, and thus there is no guarantee that growth will continue indefinitely. The rapid progress made over the past 250 years could well turn out to be a unique episode in human history. One-time-only inventions."
Robert Gordon, The Rise and Fall of American Growth:
U.S. Standard of Living Since the Civil War

SIX ECONOMIC HEADWINDS DROP GDP GROWTH BACK TO 1750

Warning: Technology's shining star is dropping out of orbit, falling, burning. Why? Cumulative innovation misfires? No. The world is poised at a historic turning point, a paradigm shift, launching a new world order. Why? Silicon Valley's famed "disruptive innovation" is backfiring.

Yes, Silicon Valley's reboot of Joseph Schumpeter's famous "Creative Destruction" economic theory from his 1942 classic *Capitalism, Socialism, and Democracy* is igniting new fears that today's mega-glitches in technology are not just economy-killers but warnings that our world is transitioning into a new order, painfully reinforcing economist Robert Gordon's prediction that America's economic growth will collapse to under 1% in this century, do in a large part to the nearsightedness of Silicon Valley's culture.

The end of technology's long run as capitalism's designated "solution to all problems" is neigh. Check out the power in the current wave of explosive failures ... stocks hit hard, a market "rout" said the *Wall Street Journal* ... "tech stocks scrape bottom," warned *USA Today* ... "optimism ebbs ... Alibaba, Qualcomm, Intel, AMD hit lowest prices in year." ... Worse, China's once-hot IPO market lost a third of its value in a month even as its government tried to stop the hemorrhaging ... now China "looks awfully like 1929, 2000."

More tech backfires: Exchange trading imploded as the longest NYSE closure in history was set off by a technology malfunction ... same day a technology glitch forced United Airlines to ground thousands of flights ... and as if a dark conspiracy were afoot, the Journal's technology overloaded, choked ... recall the FBI's recent report that millions of government employee background records had been hacked ... and punctuating the scary drama, last month right after launch, we watched the spectacular explosion of hi-tech guru Elon Musk's SpaceX Rocket.

AMERICA PRIMED TO REPEAT 1929, 2008, THANK SILICON VALLEY

Plus, it's no secret that the economy is historically weak, and the stock market is way overdue for a major correction, as in 2000, again in 2008. With tech stars dropping out of the sky. China's IPO winners fading. United Air grounded. NYSE dead in the water. The bad news is just piling up for technology. Yes, something's wrong, technology's at a historic turning point. With more bad news ahead.

Get it? Technology is looking "awfully like 1929, 2000," for everything, everywhere, the global economy, world stock exchanges, our portfolios. So, yes, this drama goes far deeper than a random hodgepodge of unrelated tech glitches. They're also not an attack of cyber-warriors from the future, Terminator robots returning to take over human civilization, the kind of drama AI futurist Ray Kurzweil predicted in *The Singularity is Near.* Nor is all this drama just temporary collateral of competing Chinese neo-capitalists vs. Silicon Valley's techno-capitalists, Jack Ma vs. Marc Andreessen.

Remember, today's six-year bull market, long propped up by the Fed's destructive cheap money policies, may be making a final push to the peak, before rolling back down, like Sisyphus' gigantic boulder crashing into our economy.

You are in a classic sequel of the soap opera of 2000 ... first a long exciting bull ... big bubble ... record peaks ... then pop goes the weasels ... a costly collapse ... later repeating the great market drama preceding the historic 2008 presidential election ... where technology won for Obama, defeated the McCain-Palin ticket, then another sequel and defeat of the GOP's Romney-Ryan ticket in 2012. And now technology's ticking time bomb looks ready to do it again in 2016.

AMERICA'S GDP COLLAPSING TO PRE-INDUSTRIAL AGE

What's really happening is far more important than all the political bickering, even the pope's sideshow. Silicon Valley is fighting a no-win battle against six fatal headwinds — cultural, political and ideological megatrends now hard-wired deep in America's collective conscience.

These headwinds are making it virtually impossible even for our best and brightest technology and science geniuses to save America from becoming a second-rate superpower in the no-growth economy that we ourselves are creating with our rigid capitalist ideology.

The reason? High-tech solutions are insufficient to prevent America's economic growth collapsing from a peak near 3.4% GDP, a cycle that began around 1750 with the Industrial Revolution and ended a generation ago with Reagan. Truth is, America's GDP is collapsing, predicted to sink into a no-growth 0.2% GDP this century—the ancient level common across the planet for centuries prior to 1750.

Yes, that's the clear message in economist Robert Gordon's must-read National Bureau of Economic Research paper, "Is U.S. Economic Growth Over?" His paper is guaranteed to push investors into total denial if not cardiac arrest, especially hard-core Silicon Valley fans who are absolutely convinced that American technology is the miracle worker that can solve all problems and is destined to save the world.

Gordon's challenge is a warning: "There was virtually no growth before 1750, and thus there is no guarantee that growth will continue indefinitely." Rather he says, "the rapid progress made over the past 250 years could well turn out to be a unique episode in human history," a collection of "one-time-only inventions." Sorry folks, but Silicon Valley cannot create a new Industrial Revolution, GDP will drop significantly. Yes, sell signal for investors.

Reread that warning: Gordon is clear, the Industrial Revolution Era is dead and gone, never to be repeated. No paradigm-shifting technologies to ignite an agricultural revolution and feed the 10 billion people on Planet Earth in 2050. Instead, just more video games, virtual reality, chat groups, digital watches and updated versions of consumer toys and consumer products, and scarce resources as buying power and food supplies disappear.

6 REASONS SILICON VALLEY CAN'T STOP ECONOMIC DECLINE

In his NBER forecast of America's future GDP growth, Gordon admits starting with a couple major biases that favored a positive, optimistic outcome: First, he admits he's "pretending that the financial crisis did not happen." Second, he admits he's making a "heroic assumption that another invention with the same productivity impact of the Internet revolution is about to appear on the near-term horizon. Thus, our starting point is quite optimistic."

Gordon's economic forecast begins with America's average GDP growth rate at 1.8% from 1987-2007. From there, Gordon's work is an "exercise in subtraction" with each of the following six headwinds reducing America's future GDP growth by percentage points, ultimately driving America's future GDP growth from 1.8% down to 0.2% by the end of the century.

Yes, that's right back to where our growth rate was before 1750, before the Industrial Revolution. Gordon's NBER paper is a must-read. Here is a summary of the six headwinds from *Bloomberg Markets:*

1. DEMOGRAPHICS

"As more and more U.S. baby boomers retire, the number of hours worked per person declines, and so does the growth in GDP per capita. (Drop down from 1.8% to 1.6%)

2. STAGNANT EDUCATIONAL ATTAINMENT

"The U.S Lags behind other advanced industrial economies in reading, math and science. (GDP growth drops more, to 1.4%)

3. RISING INCOME INEQUALITY

"From 1993 to 2008, the wealthiest 1% captured 52% of inflation-adjusted income gains." (GDP falls more, to 0.9%)

4. GLOBALIZATION & INFORMATION TECHNOLOGY

"More and more skilled jobs in the U.S. are being automated or are shifting to low-wage countries. (Drops down to 0.7% GDP)

5. ENERGY AND ENVIRONMENT

"Possible U.S. efforts to combat global warming, such as a carbon tax, act as a drag on economic growth. (America's GDP now falls further to 0.5%)

6. MASSIVE HOUSEHOLD & GOVERNMENT DEBT

"Spending money on debt repayments in the U.S. reduces funds available for productive economic activity. (GDP falls deep, to 0.2%)

Robert Gordon's closes his NBER challenge on a lighter note: "There are more than enough provocative ideas in this article, but I conclude with another. My guess is that a Canadian or Swedish economist looking at the past and future of his or her country would not be nearly so alarmed. And why not? What are the differences in environment, resources, legacy, history, policies and culture that create their relative optimism? Experts in several other countries are encouraged to contribute their own reactions to this diagnosis of the successful 'American century' and the "possibility that future economic growth may gradually sputter out."
7.15.15

ECONOMICS 101
MASTERY OF ILLUSION

*"To put it bluntly, the
discipline of economics has yet to get
over its childish passion for mathematics and for
purely theoretical and often highly ideological speculation,
at the expense of historical research and collaboration
with the other social sciences." Thomas Piketty,
Capital in The Twenty-First Century*

MYTH OF 'PERPETUAL GROWTH' KILLING AMERICA'S FUTURE

Everything you know about economics is wrong. Dead wrong. Everything. Why? Because underpinning the conclusions of today's economists is a bizarre fantasy that distorts all predictions, all forecasts, all projections, all guesstimates. Today's economic conclusions are no more real than a summer blockbuster movie, *The Avengers, X-Men, Transformers.* The difference is that the economic profession is a threat to America's future, not media entertainment. Economic dogma is on track to destroy the world with its misleading ideological assumptions.

Why? Because all economics is based on a bizarre "myth of perpetual growth." Yes, all theories and business plans based on long-term growth forecasts are distortions, usually designed after the fact to support planned business and political projects. Economists are more like salesmen who rely on a set of fictions, fantasies and irrational forecasts that emanate from the illusory magical mantra of Perpetual Growth that goes untested year after year.

ECONOMISTS & CONSUMERS LOST IN SHORT-TERM THINKING

And yet they are used to manipulate the public into a set of policies and decisions that are leading the U.S. and the world economies down a path of unsustainable globalization and GDP growth assumptions will eventually self-destruct the planet.

Yes, economists are addicted to this ideology. Trapped deep in their denial, can't see the problem, or admit it, or if they do, they are unable to stop themselves, see past their own myopic worldview. They're mercenaries working for capitalists who pay their salaries and expect them to support the capitalist's bizarre Myth of Perpetual Growth.

Worse, consumers also bought into the myth. Yes, you believe everything you learned in college about economic theories, all the textbooks, everything you read in the daily press, the government reports, all those Wall Street analysts' predictions relying on studies prepared by economists with credentials.

But everything you think you know about economics … is wrong. Dead wrong! And until economics acknowledge this failure, this missing link, the discipline of economics is driving our world down a self-destruct path.

Why? The "dismal science" of economics is not really science. Yes, it looks scientific with all the fancy math algorithms, cloud data mining, and computer models that economists use, but all that's just window dressing to make the economist look scientific, appear rational.

They're not. Their conclusions are pre-ordained, fabricated, based on their biases, personal ideologies and whatever their employer wants to prove to manipulate consumers, voters or investors to buy what they're selling.

WHAT'S AN ECONOMIST WITH A PREDICTION? WRONG!

Don't believe me? Go look at *USA Today's* quarterly surveys of 50 economists projections of GDP growth. Invariably off by a large margin. Same with *Barron's* "Big Money Poll." In past reviews we've seen a wide gap in the forecasts of the bulls and bears.

Bottom line: No matter what, you cannot trust the predictions of any economist. Ever. Be forever vigilant: Several years ago, famous *BusinessWeek* editorial headlined: "What Do You Call an Economist with a Prediction? Wrong." Unfortunately, we live in a world of capitalists who manipulate the public with the Grand Myth of Perpetual Growth, of endless growth, ad infinitum, forever, till the end of time.

Driving the economists' Grand Myth is population growth. Population is the pervasive independent variable driving their equation. Population growth drives all other derivative projections, forecasts and predictions. All GDP growth, all growth in profits and income, all wealth growth, all production growth, everything. All our assumptions of growth fit into the overall left-brain, logical, myopic mind-set of western leaders, of all the corporate CEOs, Wall Street bankers and all government policymakers running America and the world. We're living the lie.

But just because a large group collectively believes in something doesn't make it true. Perpetual growth is still a myth no matter how many economists, CEOs, bankers and politicians believe it, an illusion trapped in the brains of all these irrational, biased and uncritical folks.

NO-WIN SCENARIO: DAMNED IF WE GROW, DAMNED IF WE DON'T

Capitalism itself is at a crossroads. Perpetual growth is capitalism's sacred cow, a fantasy, a "grow or die" theory that doesn't work anymore. With us since 1776, it's being challenged by a "new god of reality" that's flashing warnings of an emerging new reality from critics, contrarians and eco-economists. This war is pitting old and new economists:

- **Grow <u>OR</u> Die. Traditional Classical Economists (Pro-Capitalism):**
  We're told we need 3% annual GDP growth to support the next batch of 100 million Americans. We believe that on faith. Go Buy stuff. Go shopping. New jobs fuel more growth. We're out of control. Exploding growth fuels demands as the rest of the world adds another three billion new humans, all chasing their version of the "American dream."

- **Grow <u>AND</u> Die. New Eco-Economists (Environmentalists):**
  They see Big Oil's destruction of our coastal economies, the rape of coal mountains, the unintended consequences of power and chemical plants spewing carbon emissions and they ask: "When will economists, politicians and corporate leaders stop pretending Earth's resources are infinitely renewable?"

We're damned if we grow, damned if we don't. Our world is at a crossroads, facing a dilemma, confronting this fundamental no-win scenario, because the "Myth of Perpetual Growth" is essential to justify supporting the three billion global population explosion coming in the next generation, by 2050. Yes, the coming population explosion is wasting our planet's limited supply of non-renewable natural resources. The growth is ultimately suicidal, will eventually destroy our civilization.

IN FUTURE, ECONOMISTS FACING NO-GROWTH GDP REALITY

But can economists change as long as they're hired hands, mercenaries employed by Perpetual Growth Capitalists? No. It will take a new mind-set. The difference between the mind-set of traditional economists and the new eco-economists is simple: Traditional economists think short-term, react short-term, pursue short-term goals. The new eco-economist thinks long-term. This may seem overly simplistic, but fits reality. Here's why:

- <u>Traditional/Classical Economists — Short-Term Thinkers:</u>
  Traditional economists are employees and consultants for organizations with short-term views, banks, big corporations, institutional investors, think-tanks, government. Population growth means new customers. They all think in lock-step, driven by daily closing prices, quarterly earnings, annual bonuses. Short-term business cycles are more important than what happens a decade in the future. Their brains are convinced: If we don't survive now, the long-term is irrelevant.

- <u>New Environmental Economists — Long-Term Thinkers:</u>
  New eco-economists see, think and plan for the long-term. They know the classical economists and capitalists thinking is setting America up for more, bigger catastrophes than 1929 or the 2008 meltdown. The "Avatar" film is a perfect metaphor: Soon capitalists will exhaust Earth's resources, force us to invade distant planets, search for new resources.

Actually, something more immediate will force change much sooner. And you are not going to like it: United Nations, World Bank and Bush Pentagon studies all predict population growth (the main driver of all economic growth) will create unsustainable natural-resources demands beginning as early as 2020 with global population exploding from seven to ten billion by 2050. And bringing with it, mass unemployment, a no-growth economy and Depression Era austerity.

6.12.12

SILICON VALLEY

DINOSAURS

"Even with the most optimistic set of assumptions, the ending of deforestation, a halving of emissions associated with food production, global emissions peaking in 2020 and then falling by 3% a year for a few decades—we have no chance of preventing emissions rising well above a number of critical tipping points that will spark uncontrollable climate change." Clive Hamilton, Requiem for a Species: Why We Resist the Truth about Climate Change

HOW 'X-PRIZE' BILLIONAIRES AVOID SOLVING 'BIG PROBLEMS'

Yes, Silicon Valley's X-Prize billionaires are failing their mission: "Making the Impossible Possible." They need at least $10 trillion to do the job, save the world, and a far bigger, better vision, as you'll see in our proposed new series of a dozen new X-Prizes that deal with the real hard issues facing our world.

Their leader is Dr. Peter Diamandis, co-author of *Abundance: The Future is Better Than You Think.* He is the CEO driving today's X-Prize Foundation and also the Chairman of Singularity University whose mission is: "Solving Humanities Grand Challenges With Accelerating Technologies." Well, he'd better step up his game.

Today's X-Prize backers resemble slow-motion dinosaurs when you think of the accelerating killer challenges facing the real world in the next generation, Planet-Earth's really "Big Problems." At best, the top four X-Prizes are feel-good ego trips for billionaires: Branson's space tourism airplane? Another low-emissions electric car? New lunar lander? Another oil spill clean-up system? Big press. Trophies for billionaires. But all marginal ideas. None big enough to save the world from itself and the "catastrophic ending" physicist Stephen Hawking predicts for Planet-Earth.

Forget the X-Prizes, what our world needs is a newer, bigger, faster 10X-Prize to tackle the real challenges. And we must start immediately, time's short. Clive Hamilton, a professor of economics and public ethics at the Centre for Applied Philosophy and Public Ethics in Australia, details the challenge in his *Requiem for a Species: Why We Resist the Truth about Climate Change.* Hamilton warns, we're our own worst enemy:

"Even with the most optimistic set of assumptions — the ending of deforestation, a halving of emissions associated with food production, global emissions peaking in 2020 and then falling by 3% a year for a few decades — we have no chance of preventing emissions rising well above a number of critical tipping points that will spark uncontrollable climate change." Game over for the UN Paris Accord?

'FUTURE IS BETTER THAN YOU THINK' (IF YOU'RE IN DENIAL!)

Hamilton warns: "The Earth's climate would enter a chaotic era lasting thousands of years before natural processes eventually establish some sort of equilibrium. Whether human beings would still be a force on the planet, or even survive, is a moot point. One thing seems certain: there will be far fewer of us."

No wonder most humans are in denial, including existing X-Prize sponsors and Singularity grads, because in denial or not global population keeps marching relentlessly from today's seven billion to the UN's predicted ten billion by 2050.

Get it? Today's X-Prize is not big enough to "save the world." But a new 10X-Prize might use a totally integrated model to accelerate the tasks. For several years we've been working with a model for innovation and investing based on the 12-part formula in Jared Diamond's *Collapse: How Societies Choose to Fail or Succeed,* 12 macroeconomic sectors that throughout history have explained why civilizations survive or self-destruct.

Diamond warns that as global population and lifestyle demands increase, "more people require more food, space, water, energy, and other resources." And that triggers more wars, famine, pandemics, no-growth economics and political miscalculations. The pressure builds as all nations begin consuming resources at same rate as America, 32 times more resources, dumping 32 times more waste. We'd need six Earths to survive.

12 'SUSTAINABLE PLANET' X-PRIZES FOR SILICON VALLEY

So, let's look to the future and see how Diamond's 12 macro-trends model might help us build a sustainable planet for 10 billion people, by setting in motion the new X-Prizes before 2020, with the two biggest prizes right up front:

Goal No. 1: New X-Prize for Sustainable Population Control

A *Scientific American* special calls population "the most overlooked and essential strategy for achieving long-term balance with the environment." Time to get past Malthus myths. In our 12-factor "Save the Earth" equation, population is the independent variable driving all others to the edge of disaster. Worse, experts challenge the numbers. Bill Gates says 8.3 billion is too high. Jeff Sachs, head of Columbia University's Earth Institute says even 5 billion is unsustainable. Paradoxically, Gates is working on population control initiatives in family planning, vaccines, contraception, all of which increase population growth.

Goal No. 2: New X-Prize to Reduce Out-of-Control Consumption

Even if population leveled off, people everywhere have their own version of the American Dream. China's President Xi Jinping underscored all this in speeches about the new 'China Dream.' China's already a military powerhouse and as Nobel Economist Robert Fogel predicts, China's economy will rise to $123 trillion by 2040, three times America's, with both of us lost in out-of-control consumerism.

Goal No. 3: New X-Prize for Abundant Global Water Systems

So essential to life: agriculture, drinking, industry, transportation. Water is more valuable than fuel for many. Fortune says, "Water is the new gold in the 21st century," generated over a half trillion dollars revenue in 2010, will soon "trade like oil futures." Already, one billion "lack access to clean drinking water" across the planet. We need massive water innovations: in purification, desalination, bottled water. New strategies like Matt Damon's Water.org, Richard Branson's Carbon War Room.

Goal No. 4: New X-Prize for Abundant World Food Production

"If you want to become rich, become a farmer," says Jim Rogers, in "Hot Commodities." Yet a billion farmers live on two dollars a day. Most are subsistence farmers. Half are underfed. Jeremy Grantham manages $100 billion, warns the planet can't feed 10 billion. Incentivize Rogers, Grantham, Cargill, Monsanto and others in a bold plan to innovate solutions for $10 trillion in private investments. This challenge is bigger more essential than Kennedy's moon shot and the Marshall Plan.

Goal No. 5: New X-Prize for Abundant Farm Land Partnerships

Rich nations like China and the Saudis are buying huge agri-land assets from poorer nations, stockpiling for the food needed for their own future population growth. But both seller and buyer need more food. There's a rush for 25% returns. Grain.org listed 416 global deals in 66 counties for 85 million acres. Buyer and seller nations must partner to eat, or revolt.

Goal No. 6: New X-Prize for Abundant Rain Forests Strategies

China and India are planning 500 new cities. Half the world's rain forests already have been converted. A quarter more by 2050. Soils are "carried away by water and wind erosion at rates between 10 to 40 times" soil formation. In forests the rate is over 500 times due to vastly increasing megafires. We need new solutions for erosion, fires, rain forest deforestation.

Goal No. 7: New X-Prize for Sustainable Fossil Fuels Strategies

Pimco's Bill Gross predicts a "significant break" in the world's "growth pattern." He says we're past the "peak oil" tipping point. His "New Normal" predicts a decline in consumer shopping, slow-growth economies, static corporate profits, while population just keeps growing. Is fracking the solution? Tar sand? Long-term risky? Get Exxon-Mobil and Bill McKibben 350.org working together on new solutions for sustainable energy policies.

Goal No. 8: New X-Prize for Abundant Solar Power Strategies

Sunlight is not unlimited. We could max out by 2050. Are fuel cells and lithium batteries enough? When Mars Rover shut down, Silicon Valley "privatized" Mars engineers to build new rockets and robots for mining energy on 10,000 asteroids worth trillions. But plans will take years before the first asteroid mining. Wake up. All 10X-Prizes will take years to plan, develop. No more delays, population won't stop racing to 10 billion.

Goal No. 9: New X-Prize for Alternative Energy Strategies

In *The Quest,* Daniel Yergin says alternative energies will remain a niche market for decades. Fossil fuels will provide 80% of total need in 2050 agrees Foreign Policy magazine. Biofuels, solar, wind and nuclear may not be a "major ticket," but with America spending a trillion annually on energy, the 20% from alternatives needs more incentives. McKibben warns: Using up all fossil reserves increases global warming.

Goal No. 10: New X-Prize for Minerals Mining Sustainability

Diamond says big plans often have unintended consequences. Mining increases GDP, but also the dumping of toxins that break down slowly in air, soil, water. Balance is essential, between insecticides, pesticides, herbicides, detergents, plastics. Innovative management strategies are needed to balance short-term economic growth with long-run public needs.

Goal No. 11: New X-Prize for in Sustainable Climate Control Regulations

Yes, human activities, autos, manufacturing produce carbon dioxide that escapes into the atmosphere destroying the protective ozone, absorbing solar energy. Innovative, risky geoengineering plans to develop and explore space-rockets technology to block harmful sun rays may backfire. Time for new strategies.

Goal No. 12: New X-Prize for Abundant Species Diversity on Planet

Transferring species to new lands has negative consequences, "preying on, parasitizing, infecting or outcompeting" native animals and plants lacking evolutionary resistance. New diseases infect native species. What's happening to bees and bats? Diamond warns, "At present rates, a large percent of the rest will disappear in half a century." We need new ideas on controlling infectious diseases and transfer of species to new environments.

Warning: This is not just another PR 'wish-list' for billionaires. These 12 goals are urgent, essential for the survival of our civilization. If world leaders fail to act, if Team Silicon Valley doesn't fund $50 trillion for 12 New X-Prizes like these, then Hamilton's warning is virtually certain to become reality: Earth's climate will "enter a chaotic era lasting thousands of years before natural processes eventually establish some sort of equilibrium. Whether human beings would still be a force on the planet, or even survive, one thing seems certain: there will be far fewer of us."

5.20.13

WWIII

RESOURCES WARS KILLING PLANET

"The world is facing an unprecedented crisis of resource depletion that goes beyond 'Peak Oil.' With all of the planet's easily accessible resource deposits rapidly approaching exhaustion, the desperate hunt for supplies has become a frenzy of extreme exploration, as governments and corporations rush to stake their claims in areas previously considered too dangerous or remote." Michael Klare, The Race for What's Left: The Global Scramble for the World's Last Resources

WWIII: MORE RESOURCES WARS KILLING AMERICAN DREAM

Yes, we are already fighting a World War: The Great Resources War to End All Wars. We've heard it before. Remember WWI, historians called it "The War to End All Wars." With 37 million casualties. WWII was even bigger, 60 million. Will WWIII finally end all war? With billions of casualties? And yes, end life on our planet? One thing for sure, it's already ending the Great American Dream.

Fasten your seat belts, soon we'll all be shocked out of denial. By sudden unpredictable catastrophic 'Black Swan' events. A global wake-up call will trigger the Pentagon's prediction in *Fortune* a decade ago at the launch of the Iraq War: "By 2020 ... an ancient pattern of desperate, all-out wars over food, water, and energy supplies is emerging ... warfare defining human life."

That's also the clear message in *The Race for What's Left: The Global Scramble for the World's Last Resources,* the latest book by noted international security expert Michael Klare. Earlier, about the same time as the Pentagon's prediction, Klare published his classic, *Resource Wars: The New Landscape of Global Conflict,* a look ahead to a world that he now hopes will not "end in war, widespread starvation, or a massive environmental catastrophe." Although they are "the probable results of persisting in the race for what's left." Unfortunately, hope can't trump reality in today's frantic race for what little is left.

We need men who pull no punches in describing what's dead ahead, whether labeling it "Resource Wars" or even WWIV, The Great Commodities War That Can End Everything." Klare does just that with this warning:

"It is true that eliminating our dependence on fossil fuels and other finite materials cannot be accomplished overnight — our current reliance on them is just too great," warns Klare, well aware the forces of capitalism are trapped in denial, cannot see the dangers dead ahead, focus only on getting richer no matter the consequences to the planet.

"But no matter how much corporate or government officials wish to deny it, there is not nearly enough non-renewable resources on this planet to perpetually satisfy the growing needs of a ballooning world population."

WORLD NATIONS ALREADY AT WAR FOR MORE RESOURCES

Even worse, in today's world run by climate-denying billionaires, Klare warns "existing modes of production are causing unacceptable damage to the global environment. Eventually continuing with current industrial practices will simply prove impossible. And precisely because implementing a whole new industrial order will be a lengthy task, any delay in beginning that work will prove costly, as resources keep dwindling and their prices continue to rise."

If there is a race, it's a downhill race to WWIII: The Great Commodity Wars. The world's great powers are accelerating war preparations — yes, they are in the early logistical build-up stage, amassing the resources and arms to send troops into battle.

And they're doing it in a world lost in denial, sinking deeper into a collective conscience that pretends our problems will be solved by the magic of free-market capitalism, unwilling to admit it not only no longer exists, it has morphed into an anarchy controlled by a bizarre conspiracy of Super Rich narcissists.

GLOBAL RESOURCES WARS: DEPLETION, AUSTERITY, POVERTY

Yes, the planet is at a historic turning point. You must plan for black swans, earth-shaking wake-up calls — a perfect storm of global wars, mass starvation, pandemics, environmental catastrophes.

The critical mass is building. We're just not listening, especially conservative politicians, Wall Street CEOs and the Super Rich, who dismiss the warnings of men like environmentalist Bill McKibben, money manager Jeremy Grantham, anthropologist Jared Diamond and global security expert Michael Klare all warning us to wake up, before it's too late to react, let alone plan.

Listen to the warnings: "The world is facing an unprecedented crisis of resource depletion — a crisis that goes beyond 'peak oil' to encompass shortages of coal and uranium, copper and lithium, water and arable land. With all of the planet's easily accessible resource deposits rapidly approaching exhaustion, the desperate hunt for supplies has become a frenzy of extreme exploration, as governments and corporations rush to stake their claims in areas previously considered too dangerous or remote."

GRAB WHAT'S LEFT ... TILL NOTHING'S LEFT... FOR NOBODY

Klare opens on a fascinating replay of Russia's 2007 risky deployment of a mini-submarine using a robotic arm to plant a titanium flag deep under the polar ice cap, two and a half miles below the surface of the North Pole. Why?

Forget national pride. In recent years as climate change warms this "frozen wasteland," Russia, as well as Canada, the U.S. and other nations are laying a claim to long-ignored "vast deposits of oil, natural gas and valuable minerals."

Faced with an impossible equation — out-of-control global population growth plus rapid depletion of nonrenewable resources equals mega-catastrophes — the big players are all selfishly grabbing and hoarding scarce commodities ... like desperate banana republic dictators as the entire world sinks into pure anarchy, scrambling for a share of what little's left, until nothing is left for anyone.

13 REASONS WHY THIS WAR IS SO DIFFERENT, SO CATASTROPHIC

This time the challenges the world is facing really are very different from any prior time in history, warns Klare: "While the current assault on remote resource frontiers bears some similarities to the historical exploration of undeveloped territories," such as the Roman Empire's expansion, today's global threats are "in many important ways different from anything that has come before."

Why? Because "never before have we seen the same combination of factors that confronts us today." Here are the five biggest reasons the next few decades are so crucial to the survival of the planet and our civilization:

1. Nonrenewable commodities, rapidly, permanently disappearing.
2. There are no "new frontiers" to open, existing reserves gone forever.
3. Population, "sudden emergence of rapacious new consumers."
4. Economic, environmental costs limit future explorations.
5. Climate change is having "devastating" consequences on energy.

Klare adds that in "many cases, the commodities procured will represent the final supplies of their type." Yes, "the race we are on today is the last of its kind that we are likely to undertake." Seven other factors are reviewed or come to mind that definitely are risk factors that increase the probability of massive global catastrophes:

6. Rapid rise of powerful resources competitors, China, Africa, Saudis
7. New warrior mind-set willing to fight for new territories and borders.
8. Conservative strategies hang on to existing industrial methods rather
9. than develop new technologies and innovative alternatives.
10. Political will lacking to invest government funds to prime innovation.
11. The time needed to prepare for known threats is rapidly vanishing.
12. America's rapidly morphing from a democracy to a plutocracy.

13. Failure to grasp that this new era of "peak everything" means that
14. the lack of resources will increase scarcity and austerity for all
     nations.

And finally, there is a total failure to acknowledge and plan any kind of population controls. Just the opposite, we are even denying birth control while encouraging population expansion feeding into the myth of perpetual economic growth.

THE GREAT RESOURCES WARS ARE BEING FOUGHT TODAY

Soon, even the myopic dinosaurs in the oil, coal and fossil-fuels industries, the guys who have been bragging about having 200 or more years of reserves, will be hit with a catastrophic wake-up call, as these risk factors balloon to critical mass and a flash point — fueled by commodity wars, pandemics, global starvation, environmental crises, skyrocketing commodity prices and accelerating population growth.

But by then, as Klare and others like him warn, it will be too late for the fossil-fuel dinosaurs ... and for everyone else. Whether you're a hard-line climate-denying billionaire capitalist or a liberal-leaning environmentalist, you need to read Klare's new *Race For What's Left: The Global Scramble for the World's Last Resources.* Or as I often call it, either, *The New Era of Depletion, Austerity and Collapse ...* or *WWIV: The Great Commodities War-to-End-All-Wars ...* and end sustainability for our world.
8.7.12

QUESTION #3:

"WHAT'S 'NEXT' FOR 10 BILLION HUMANS LIVING IN 2047?"

2047

THE POINT OF NO RETURN

*"Temperatures in New York are increasing, after 2047 they won't
return to the historical average of the past one and half centuries,"
The study's director warns, "within my generation, whatever
climate we were used to will be a thing of the past, for some
as early as 2020." Bloomberg Business News*

SOON WORLD'S CITIES TOO HOT, UNBEARABLE FOR LIVING

Yes, the conservative right agenda is clear: Global warming is a massive liberal hoax. A GOP win big for Big Oil, capitalist billionaires and climate-science deniers confirms the worst fears of environmentalists worldwide: The planet's high-speed crash-and-burn cycle will accelerate. The message to cities across America and worldwide: Burn baby burn!

Here's why: The GOP agenda demands more energy use. Pass Keystone XL. Gut EPA. Export domestic oil. More deep-water fracking. Drilling in national parks. Dig more coal. Offshore drilling. Partnering with Russia in the Arctic. Curb renewable energies. Yes, our world's getting hotter than hell. Drill baby drill. Burn baby burn!

Big Oil, the GOP and their vast army of climate-science deniers obviously see things different from the Dems. Energy giants like Exxon Mobil and the Koch Bros won big betting on GOP governors, senators and representatives. Now they control the U.S. Senate, House and about two-thirds of the state governments in America, legislatures and executive branches. They own the new energy growth economy.

And yet, even if America's climate-science deniers don't believe it, others do: Planet-Earth is on a fast-track hothouse trend that NASA, the United Nations and thousands of scientists have been warning about for two decades. Today that's nowhere more obvious than in the rising temperatures that will heat and burn across major U.S. cities, making urban life intolerable.

RAPID SHIFTS: FARMS TO OVER-HEATED CITY HIGH-RISES

Yes, midway this century, around 2047, urban areas globally and across America — New York City, San Francisco, Chicago, Houston, Seattle, Los Angeles, Washington — will become so unbearably hot, uncomfortable, local economies will stagnate, and many will be unable to host the Olympics later in this century.

Midcentury, worldwide population is predicted to explode from seven billion today to 10 billion, mainly in urban areas. Today 54% live in cities. In three decades it'll be 66%, as high as 90% in some rapid growth regions in Asia and Africa. Many nations are urbanizing even faster: China's aggressively relocating 250 million farmers, building hundreds of new cities. All need food, water, lots of energy. All in one generation.

Yes, in less than three decades, around 2047, when babies born today are starting families, our world will be passing a point of no return: Our global hothouse will never return to earlier long-term temperature averages, according to a joint study by the University of Hawaii and Japan's University of the Ryukyus, which was first reviewed in the *Nature* journal.

MARS OR VENUS: SOON YOUR CITY WILL BE UNBEARABLE

Here's *Bloomberg News* summary of that study: "Temperatures in New York are increasing, and after 2047 they won't return to the historical average of the past one and half centuries," writes Alex Morales. The study director, Camilo Mora, a geographer, said: "The tropics will experience unprecedented warming 15 years earlier than the rest of the world," warning that "within my generation, whatever climate we were used to will be a thing of the past, some as early as 2020."

Yes, after 2047, the point-of-no-return average for all cities worldwide will go over a cliff, never returning to historical averages. Never. Earth will just keep heating up. Living in big cities like New York, Washington, San Francisco, will become unbearably hot, uncomfortable.

I had an early taste of it: Living in New York City. With Morgan Stanley during a blackout some decades ago. Hot summer. No electricity. Elevators dead climbed 23 flights. Bad news. Later, as a consultant in Saudi Arabia one summer I was thankful for their air-conditioning. But by 2050, energy demands will trigger constant high-rise blackouts. For many mid-day temperatures are already too hot for laborers outdoors.

POINTS OF NO RETURN: OVERHEATING IRREVERSIBLE

Here's the University joint research team's fascinating global timetable. Overall, the median year for all cities is 2047. Then, worldwide average will pass a point of no return, *never to drop back to historical averages,* with a profound impact on the next generation ... all denied in the energy policy agenda of conservative politicians:

- **2020-2029 decade:**
  The first, New Guinea's climate will "permanently enter a state never seen before, outside of the bounds of historical variability and short-term extremes. To put it simply: The coldest year in New Guinea after 2020 will be warmer than the hottest year anyone there has ever experienced." Manokwari, Indonesia, will join them as the earliest burnouts. In 2029, Jakarta and Lagos.

- **2030-2039 decade:**
  In the 2030s several mega-million cities pass into the permanent hot zone, never to return: Mexico City 2031, Bogota 2033, Mumbai 2034, Baghdad 2036, Cairo 2036, and Nairobi also in 2036.

- **2040-2049 decade:**
  In the 2040s decade these major cities enter a permanent hot zone: Tokyo 2041, Pretoria 2043; Santiago 2043, Rome 2044, Beijing 2046 and Bangkok 2046. And America joins the hot 2040s decade: Orlando 2046, New York and Washington 2047, San Francisco in 2049. So our politicians can ignore the trend for a couple more decades before they wake up, too late.

- **2050 and later to 2073:**
  Then after 2050, when U.N. demographers forecast that global population will have exploded from 7 billion today to 10 billion, several other mega-million metropolises will pass the point of no return, entering the planet's permanent hot zone: Rio de Janeiro in 2050, London 2056, Moscow 2063, Reykjavik 2066 and Anchorage in 2073. We know Alaska is already losing a lot of its ancient permafrost. Rising temperatures are turning it into an unstable mud hole.

Rex Tillerson, former Secretary of State and ExxonMobil's CEO is a climate-science denier dismissing all this, even though Exxon began researching climate warming almost fifty years ago. Still, he doesn't trust "climate models to predict the magnitude of the impact." Instead, he simply has faith that humans will "adapt to a sea-level rise." After all, throughout history, he says humans "have spent our entire existence adapting. We'll adapt." We just need a little optimism, right?

CAN PLANET-EARTH'S 10 BILLION HUMANS 'ADAPT' BY 2047?

But unfortunately, it will happen too fast to shift gears and plan ahead. America's capitalists, climate-science deniers, and Wall Street energy traders are so trapped into making money as fast as possible, they're unlikely to 'wake up' in time. What about Washington lobbyists who make hundreds of millions from Big Oil? Will passing these points of no-return wake up America's political machine that feeds off payoffs from the myopic greed of Big Oil capitalists? Hardly, their narrow vision will remain impenetrable ... till it's too late.
11.10.14

SHOCK

DOCTRINE

"What is behind the abrupt rise in climate change denial among hardcore conservatives: they have come to understand that as soon as they admit that climate change is real, they will lose the central ideological battle of our time—whether we need to plan and manage our societies to reflect our goals and values, or whether that task can be left to the magic of the market." Naomi Klein, This Changes Everything: Capitalism vs The Climate

WHY CAPITALISM IS WORLD'S BIGGEST PROBLEM, NOT SOLUTION

GOP conservatives, Big Oil, Exxon Mobil, Koch billionaires, and every other hard-right climate-science denier must love Naomi Klein's latest book, *This Changes Everything: Capitalism vs The Climate.* Why? My first response, this is old news, the title should be: *This Changes Nothing: Capitalism Still Wins, Climate is Still Losing.* Sorry folks, but the future looks brighter than ever for capitalism.

We are in the real warzone. Klein saw the world sinking deep into a "capitalism vs. climate" global conflict since well before her last book, *The Shock Doctrine: Rise of Disaster Capitalism,* a historical survey of the conservative revolution launched after WWII by Nobel Economist Milton Friedman and Ayn Rand, patron saint of capitalism. That revolution sunk its roots deeper into our culture and history under the leadership of President Ronald Reagan and the 18-year reign of Fed Chairman Alan Greenspan.

In our early review of *The Shock Doctrine,* we called it "one of the best economics books of the new 21st century." Since then the zeitgeist has moved past *Shock Doctrine.* Americans sense we are sinking deeper into chaos, as climate issues accelerate, creating deep ideological divisions. Klein detailed how conservatives emerged so successful in their power grab the past generation since the Reagan revolution. In fact, paradoxically, *Shock Doctrine* has motivated conservatives to grab for more and more power.

Yes, conservative strategies are working. And Klein's new book is guaranteed to further emboldened conservatives to build on their power base ... further accelerating America's downward spiral ... as capitalism keeps widening the global inequality gap ... as the 67 billionaires who now own half the world will keep grabbing more ... as Credit Suisse Bank's prediction that by 2100 eleven trillionaire families will rule the planet seems more credible ... as economist Thomas Piketty's warning that capitalism has become so powerful it is unstoppable, that it will continue widening the inequality gap, in spite of Pope Francis warning that inequality as the "root cause of *all* the world's problems."

Klein's new book was to be a game-changer. Now it may be old news. Changed nothing. Capitalism keeps winning. Planet-Earth keeps losing. The global capitalist ideology truly is the global zeitgeist, pouring more fuel on the fire, accelerating a downward spiral.

NEW STRATEGY: REVOLUTIONS? OR ALREADY TOO LATE?

Klein's real big strategy: Start a revolution. Forget piecemeal government policies. Band-Aids aren't working. Not for liberals. If you want change — real change — start a real revolution, a grassroots revolution, depose any capitalist rulers. But be prepared, you may not like the consequences. Revolutions have a bad habit of taking down economies, making life worse for everybody, not just the rich and powerful, as in the 1789 French Revolution.

Sorry, but it may already be too late. Time's running out. Unfortunately, Klein's new book may be too late to be a game-changer. The forces against climate change — *Big Oil, GOP, conservative billionaires, fundamental-science deniers* — have become far too powerful, too politically polarized, too rigid, too uncompromising and, yes, too rich, despite the UN-IPCC's Paris Agreement.

The truth is, most Americans don't want change. We're a capitalist nation. Gallup polls tell us 76% of America say climate change is not a national priority. We live in an automobile culture. We need gas pumps. Yes, we need Big Oil. The American Dream is programmed ... against climate change and global warming.

So unfortunately for aspiring game-changers like Klein, the climate-science-denying conspirators and most Americans actually love the way their collective brain works. Thousands of books have been written by the 2,500 scientists on the UN International Panel for Climate Change committees since 1989, but they did not stop the march of global warming, just the opposite, hardened the opposition, encouraged the conservative revolution.

Yes, capitalism itself will eventually self-destruct, because competition for new markets and ever-scarcer resources, accelerating global warming and climate disasters will ultimately kill off much of the human race.

And it's not just that the deniers like Big Oil who resist solely because they know carbon emissions regulations and taxes will upset their economic model. Nor because the capitalist brain is hard-wired on short-term profits and is incapable of balancing today's profits against a longer-range future, discounting future costs to nothing, leaving problems for future generations to figure out and expense. All that's ancillary.

SOLVE CAPITALISM PROBLEM WITH CAPITALIST THINKING?

The real reason? Capitalists believe capitalism is not the problem—capitalism is the solution. To everything. That core belief means their brains can't grasp Albert Einstein's warning, that "we can't solve problems by using the same kind of thinking we used when we created them."

Einstein's warning was reinforced in **David Owen's** recent, *The Conundrum: How Scientific Innovation, Increased Efficiency, and Good Intentions Can Make Our Energy and Climate Problems Worse.* Why? If any new technology increases the consumption of scarce resources, it's making matters worse for an over-populating planet. So, capitalists and their solutions will just keep making matters worse for everyone this century. Until a catastrophe, a revolution self-destructs capitalism, and probably the rest of us.

The problems created by capitalism in the last couple generations have gotten so out of control they've morphed into an unstoppable conspiracy of Big Oil, GOP and hard-right conservatives that is hell-bent on global dominance ... even gambling the future of the human race and our civilization on their blind conviction that capitalism is the solution to all problems.

Bottom line: Ironically, the conservative revolutionaries must love Klein's new book the most. It adds richness to the history of the rapid rise to power of their conservative revolution while articulating a strategic plan for their continued rise to total world domination. They also know how to hedge a bet, even have a backup Plan B, just in case they're wrong and capitalism is not the solution to America's capitalism problem.

Remember the sage advice of Barton Biggs, long-time research director at Morgan Stanley, my old Wall Street firm. In his classic, *Wealth, War and Wisdom,* Biggs, later a very successful hedge fund manager, advised his super-rich clientele to expect the "possibility of a breakdown of the civilized infrastructure."

His advice: Buy a farm isolated in the mountains. Protect your family against the barbarians, "think Swiss Family Robinson, your safe haven must be self-sufficient, capable of growing food, well-stocked with seed, fertilizer, canned food, wine, medicine, clothes. And be ready to fire a few rounds over the approaching brigands' heads, persuade them there are easier farms to pillage."

Wake up America: the revolution is coming. Klein got it right. Your capitalist brain may not see it yet, may not till too late. But prep for when you do. Learn how to farm, now.

10.8.14

THE GREATEST

HOAX

"God's still up there...climate is changing and climate has always changed and always will. The hoax is that there are some people who are so arrogant to think they are so powerful they can change climate. Man can't change climate, to think that we, human beings, would be able to change what He is doing in the climate is to me outrageous." U.S. Senator Jim Inhofe

SENATOR SAYS "GOD'S IN CHARGE," BUT WHAT IF 'HE'S' NOT?

Climate? Warming? God's in charge. Humans powerless? Why? The Bible's their authority, God has absolute control. And yes, that message is absolutely clear, says Oklahoma Senator Jim Inhofe, the powerful Republican chairman of the Senate Environment and Public Works Committee and author of *The Greatest Hoax: How the Global Warming Conspiracy Threatens Your Future:*

Inhofe says "climate is changing, and climate has always changed and always will. The hoax is that there are some people who are so arrogant to think they are so powerful they can change climate. Man can't change climate."

For years Inhofe has been preaching this biblical message: "My point is, God's still up there. The arrogance of people to think that we, human beings, would be able to change what He is doing in the climate is to me outrageous." In other words, if the planet's climate really is changing, it is not "human-caused," God's in charge, God's responsible, blame God.

We also know Senator Inhofe is not some lone voice crying in the wilderness. Millions of Americans agree with him. A recent *Washington Post's* religion column brought this point home: "People are seriously wondering whether God is punishing us with the 2016 election," wondering, is this the "end-times?"

But what if Inhofe and the evangelicals are wrong—what if God is *not* in charge? God may just be testing us, as in *The Book of Job* when His "Whirlwind" was unleashed creating death and destruction. Or he may be seeing how we do with free will, with the brains given us for managing Creation? Or what if God's more of a divine "intelligent designer," the ultimate MasterCoder, who created the universe, set it in motion, then stepped back, observing till later, keeping score till judgment day?

WE'RE SHORT-TERM THINKERS, DELAY TILL IT'S TOO LATE

Worse yet, tens of millions don't even think about God's role, or care enough to spend taxpayer dollars to solve our climate warming problems. In fact, the Yale University's "Six Americas" study, Gallup polls and other research tells us that a majority of Americans actually do agree with the good senator.

In fact, just last year an overwhelming 76% of Americans put climate change near the bottom of 15 "national problems" in Gallup polling. Yes, global warming is a problem. But not today, they'll worry about it later. Somewhere between 100 million and 225 million Americans, hard-right evangelicals, superrich capitalists, employees of the fossil-fuels corporations or, as Gallup puts it, citizens who just don't think climate is a big problem for America.

Why? America has bigger, more urgent problems: the economy, jobs, terrorism, hunger, drugs, crime, immigration, deficits, and many more. That's human nature, we are short-term thinkers, today's problems, even when later may be too late to plan and solve any climate problems.

Get it? Climate is a problem, but not a big enough problem. Someday maybe, but not today when the costs weigh heavy on other pocketbook issues. And not when we're not sure that investing big money in solutions will actually decrease global temperatures, ocean-currents pollution or carbon dioxide and not after we've already spent heavily to control fossil-fuel use in our personal autos, to increase EPA regulations on coal and to save the rain forests.

POLITICIANS FAVOR BIG OIL VS GLOBAL WARMING REGS

In addition to millions of voters, Inhofe's biblical message also has the support of the controlling majority in the U.S. Congress, specifically 169 hard-right GOP climate-science deniers who dismiss the idea that global warming is human-caused. They got the votes.

Plus, there's no doubt how they'll vote. Collectively the GOP climate-science deniers have received more than $52 million in career contributions from Big Oil, coal and fossil-fuel interests whose lobbyists demand the candidates they finance will fight to defeat or delay all efforts to tax and regulate carbon emissions.

Moreover, since all the GOP presidential candidates must toe the pro-business party line, it's a solid bet that a Republican president would guarantee that Inhofe's biblical interpretation of climate change would be enacted into federal law at least till 2020.

Get it? Climate is emerging as a defining issue for voters in America's elections, when it comes to climate change and global warming, for these millions, God is solely in charge. Period. Humans are powerless, cannot change climate. God is in charge.

Their message is quite simple: No matter what how hard we try, humans can have no impact on our climate. Nor can we make climate worse. God controls all changes, increases and decreases in global warming and changes in climate. Humans are powerless to do anything.

BIBLE AS AUTHORITY: GENESIS? JOB? REVELATIONS?

So, what's Inhofe's moral authority for trusting solely in God, insisting humans have zero impact and can do nothing about climate and global warming? Evangelicals rely on a literal interpretation: "One of my favorite Bible verses" says Inhofe, is in Genesis: "As long as the Earth remains, there will be springtime and harvest, cold and heat, winter and summer." But doesn't that give humans both the responsibility for God's Creation, and also the power to change it?

While we respect the good senator and other Republicans for their strict adherence to scriptural authority, the Bible has many other, often conflicting, moral lessons guiding us: in the stories of Noah, Joseph, Moses, Jesus and more. *The Book of Job* found me many years ago when I was challenged by events, going through another midlife crisis, lost, in the early days of recovery from heavy addictions, depressed, sinking deep into a very dark night of the soul.

One day my search led me wandering far from home, to the Mary and Joseph Retreat House in Palos Verdes California and a lengthy conversation with an elderly nun, a truly peaceful soul. Afterwards, she gave me a Bible: "Go into our gardens, sit in the sun, open it at random, read, you'll be guided."

IF GOD'S IN CONTROL, THEN WE'RE WASTING HIS CREATION

The Book of Job grabbed me, the thunder of a God and his awesome Whirlwind. I saw, knew, identified with what Job was going through. Reading The Book hit hard, triggering buried childhood traumas and times when to survive I felt spirits coming to protect me, transporting me to safety, traveling across time to other worlds, to safe places. Since then The Story of Job has been part of me and the world I see.

Job was a righteous and richest of men, a faithful servant of God. But still, He was testing Job. His family, ten children, herds of thousands, all destroyed in a great climate catastrophe by God's Whirlwind. Tested even more when his health was tortured. Then his best friends even accused him of sin. Yet he was a true and faithful servant. Yes, bad things do happen to good people.

Job speaks to me often, in many ways, not as a parable, nor a favorite story, but as a living experience, my eyes into today's world. We all have such contacts. You know what I'm talking about. We listen to that still, small voice deep in our souls. We hear, each in our own unique way, lines opened many times with guides over decades, eons.

The message is clear: Humans are being profoundly tested again. Do we do nothing? Or is it time for action, time to work with Him. Protect His Earth. Passive acceptance? Hope? Prayer? Not enough. Bill McKibben once told me, "even if you believe it's already too late, keep trying." That's what this war is all about, there is no hoax.

Yes, we are being tested again. So, whether you're the next "Job from Uz," or from Hawaii, or Singapore, or New York City, doing nothing really is not an option. Go directly to the source. You're in a partnership. With whoever your God is, the still small voice in you, your spirit, your collective consciousness. Act responsible. Work with Him. Protect His Earth ... before it really is too late ... remember, it's your Earth too.

4.22.15

NO PLAN-B?

REVOLUTIONS!

"The ideologues of rapacious capitalism, like members of a primitive cult, chant the false mantra that natural resources and expansion are infinite. It all will come down like a house of cards. Civilizations in the final stages of decay are dominated by elites out of touch with reality." Chris Hedges, Empire of Illusion

SCIENCE-DENYING CAPITALISM FUELING GLOBAL REVOLUTIONS

The Right has no "Plan B" just in case we pass the point of no return. No plans for when global warming becomes irreversible? None. Zero. Big Oil, GOP ideologues, conservative billionaires, are all absolutely certain they don't need a back-up plan. Why? Because capitalism is the solution to all of America's problems. Capitalism is the one and only solution to all problems in the entire world. So, for them, no 'Plan-B.'

But ask yourself … What if global warming really does become irreversible? And The Right is wrong, has no "Plan B," and their "Plan-A" is total denial? Then what? Too bad, once Planet-Earth passes the point of no return, it's too late for any plans, no turning around, because then climate change goes out-of-control, irreversible.

Seriously, what if later it turns out that the gang of the GOP, Big Oil's Exxon Mobil, the Koch Bros, Chamber of Commerce, the president and a lot of conservative billionaires, their army of well-financed lobbyists, climate-science deniers and senators like Oklahoma's Inhofe — really are all totally wrong? Dead wrong?

What if by 2050, in just one generation, after we add three billion more humans, what if Planet-Earth does in fact pass a point of no return and global warming becomes permanently irreversible? What if our Planet Earth can no longer cool? Can no longer revert to historic temperature patterns? What if our planet just keeps getting hotter? Wars over water, food. Turning into parched deserts? Into predictable 1,000-year dust bowls?

Then it's game-over, right? Not just for the 67 billionaires Forbes tells us already own as much as the poorest half the world, but for every last one of the 10 billion humans living on the planet in 2050.

IT'S ALREADY TOO LATE, CAPITALISM'S SELF-DESTRUCTING!

Actually ... it is already too late ... we're in denial ... it is too late, we just can't admit it! "Karl Marx had it right," said economist Nouriel Roubini in the *Wall Street Journal:* "At some point, capitalism can destroy itself. ... We thought that markets work. They are not working. What's individually rational ... is a self-destructive process," for capitalism. So, ask yourself again: What if the Right is wrong? If capitalism fails? It's too late? And unfortunately, it'll also be too late to start planning a "Plan-B."

So, listen closely, capitalists are self-destructing capitalism, sabotaging American democracy, destroying the global economy. Get it? We won't have to wait till 2100. Self-destruction is happening now, today, and will be completed in the next generation, by 2050, mid-century. Here's how Chris Hedges, Presbyterian minister, former war correspondent and author of many best-sellers, including *Empire of Illusion,* when he discussed on Truthdig.com:

> "The ideologues of rapacious capitalism, like members of a primitive cult, chant the false mantra that natural resources and expansion are infinite ... it all will come down like a house of cards. Civilizations in the final stages of decay are dominated by elites out of touch with reality." They "strain harder and harder to sustain the decadent opulence of the ruling class, even as it destroys the foundations of productivity and wealth."
> "This failure to impose limits cannibalizes natural resources and human communities. This time, the difference is that when we go the whole planet will go with us. Catastrophic climate change is inevitable ... There will soon be so much heat trapped in the atmosphere that any attempt to scale back carbon emissions will make no difference. Droughts. Floods. Heat waves. Killer hurricanes and tornados. Power outages. Freak weather. Rising sea levels. Crop destruction. Food shortages. Plagues."

"Exxon Mobil, BP and the coal and natural gas companies," warns Hedges, "will never impose rational limits on themselves. They will exploit, exploit, exploit. Collective suicide is never factored into quarterly profit reports." The Righteous Right is on a "collective suicide" trajectory, dragging the entire world over the cliff with it. No wonder Roubini says that capitalism will inevitably self-destruct. It's blind, arrogant, obsessed.

WHY CAPITALISTS CAN'T ACCEPT LIMITS ON FREE-MARKETS

"Resistance may ultimately be in vain," warns Hedges: Capitalists will self-destruct Planet Earth. Solution? Global revolutions. For "to resist is to say something about us as human beings. It keeps alive the possibility of hope, even as all empirical evidence points to inevitable destruction. It makes victory, however remote, possible. And it makes life a little more difficult for the ruling class, which satisfies the very human emotion of vengeance."

Get it? Even though we're facing "inevitable destruction," due to "collective suicide" of capitalism, it's time to go beyond passive protesting like Bill McKibben's global 350.org network of peaceful activists. It's time to wake up, and see the odds of successful revolutions are very high now: Just a few thousand disorganized billionaires versus 7.3 billion humans who will soon wake up realizing they really have nothing to lose on a planet that by 2050, as Jeremy Grantham warns, will not be able to feed 10 billion humans anyway. Solution: start the revolution, also advice in Naomi Klein's new classic, *This Changes Everything: Capitalism vs. The Climate.*

SUCCESSFUL REVOLUTIONS: FLASHPOINT, IGNITION, RAGE!

Years ago, Hedges began commenting on out-of-control capitalism and the consequences of extreme inequalities in the world. This same widening economic gap witnessed by leaders like Pope Francis, Nobel Economist Joseph Stiglitz and Tom Piketty, author of *Capital in the 21st Century.* Writing on Truthdig, Hedges wrote, "This is What Revolution Looks Like," anticipating game-changing revolutions ahead for Capitalist America. Listen:

"Welcome to the revolution." It's already happening. "Our elites have exposed their hand. They have nothing to offer. They can destroy but they cannot build. They can repress but they cannot lead. They can steal but they cannot share. They can talk but they cannot speak. They are as dead and useless to us ... They have no ideas, no plans and no vision for the future." Hedges then references historian Crane Brinton's book, *Anatomy of a Revolution,* focusing on "the preconditions for successful revolution:"

— Discontent that affects nearly all social classes ...
— Widespread feelings of entrapment and despair ...
— Unfulfilled expectations ...
— Unified solidarity in opposition to a tiny power elite ...
— Refusal by scholars and thinkers to defend the actions of the ruling class ...
— Inability of government to respond to the basic needs of citizens ...
— Steady loss of will within the power elite itself.
— Defections from the inner circle ...
— Crippling isolation leaves the power elite with no allies or outside support."

But reading these first nine "preconditions" in the context of today's America in chaos, corrupt leadership, do-nothing Washington, out-of-touch capitalists, it seemed obvious that these nine "preconditions of a revolution," by themselves aren't enough, at least not yet, more is needed. But then an alarm bell rang, loudly.

Hedges gets our attention with his tenth precondition—a "financial crisis," which should remind all Americans of the damage from the 2008 bank credit collapse that pushed Wall Street into de facto bankruptcy and cost investors over $10 trillion. So, add in the tenth "precondition, a financial crisis," and with that we have "fulfilled the preconditions." Then we have some loud responses to the big question ... No Plan B? What if the Right is wrong? Can we wait till 2020? Time for a revolution is now!
11.22.14

SIXTH SPECIES

EXTINCTION

*"Climate change is a matter of geologic time, something that the
Earth routinely does on its own without asking anyone's permission
or explaining itself" and "doesn't include the potentially catastrophic effects
on civilization in its planning. Far from being responsible for damaging the
earth's climate, civilization might not be able to forestall any of these
terrible changes once the Earth has decided to make them ...
Climate ought not to concern us too much ... because it's beyond
our power to control." Robert B. Laughlin, Nobel Physicist*

RECYCLE? HYBRIDS? SOLAR? WE CAN'T STOP EXTINCTION

"The Earth Doesn't Care If You Drive a Hybrid!" Or recycle. Or eat organic food. Or live in a greenhouse powered by clean energy, solar energy. Or squander scarce resources. Sorry folks, but Planet-Earth just doesn't care how much you waste.

Was that published in *Mother Earth News*? Satire in *The Onion*? No, it was the actual cover story in the elite *American Scholar* journal, written by Nobel Geophysicist Robert B. Laughlin of Stanford University. I bring it to your attention because in today's irrational climate science denialism arena, this may be the only corner where politicians and scientists agree.

Actually, Laughlin's message is rather conservative: "Far from being responsible for damaging the earth's climate, civilization might not be able to forestall any of these terrible changes once the earth has decided to make them ... The geologic record suggests that climate ought not to concern us too much when we're gazing into the energy future, not because it's unimportant, but because it's beyond our power to control."

Remember, "Earth didn't replace the dinosaurs after they died" during the last great species extinction, says Nobel physicist Robert B. Laughlin. She "just moved on and became something different." But so what, you say, wasn't that 65 million long years ago? It won't happen again in our lifetime?

NEW DINOSAURS: TRIGGERING OUR OWN EXTINCTION!

Unfortunately, humans are the new dinosaurs, the next species slated for extinction, warn 2,000 United Nations scientists. Soon. We're also causing the extinction, even accelerating on a new timetable. Signing our own death warrant. Not millions of years in the future, but this century. Thanks to our secret love fueling climate change with fossil fuels. Yes, we're all closet science deniers.

Here's how Laughlin put it: "Humans have already triggered the sixth great period of species extinction in Earth's history." Get it? We're to blame. We are the engine driving the next species extermination. The human race is on a suicidal run to self-destruction. We can't blame it on the great American conspiracy of climate-science deniers, Big Oil, the Koch Bros, U.S. Chamber of Commerce and Congress. It's us.

We just keep buying gas guzzlers, keep investing retirement money in Exxon Mobil, keep making more and more babies, forever in denial of the widening gap between perpetual economic growth and more babies living on a planet of rapidly diminishing resources.

Yes, humans have already triggered the sixth great species extinction. And yes, it's already in progress for this century. Why? Because we're solving the wrong problems, using the wrong data, applying the wrong formulas. Even the U.N.'s Intergovernmental Panel on Climate Change with its 2,000 elite scientists. They update us with 2,000-page technical reports, every five or six years since 1988.

WRONG PROBLEM ... BAD SOLUTION ... NO ALGORITHM

But they're solving the wrong problems. As problem solvers, the U.N.'s army of 2,500 climate scientists aren't much different than ExxonMobil's CEO Rex Tillerson. He admits climate change is real. But he believes it's just an "engineering problem and there will be an engineering solution." Tillerson also doesn't trust those "climate models to predict the magnitude of the impact." Like the U.N. scientific models. Instead, Tillerson has faith that humans will "adapt to a sea-level rise." We've "spent our entire existence adapting. We'll adapt."

So even if the U.N. has 20,000 scientists who are 100% certain that climate change will wipe human civilization off the planet like dinosaurs, never to return ... still you can bet your Big Oil retirement stock that Tillerson and every other science denier will keep fighting for free-market capitalism, subsidies and deregulation, keep investing $37 billion annually in exploration. And with their war chest of $150 billion annual profits, they can still pay off all the politicians and investors they need to make sure Big Oil keeps beating all the U.N.'s climate scientists.

BIGGEST PROBLEM: OVER-POPULATION? NO, DENIAL!

What's wrong? Everybody on Earth is in denial about our biggest problem ... population growth. Too many new babies, a net of 75 million a year. Yes, we're all closet deniers — leaders, investors, billionaires, politicians, the 99%, everybody, even Bill McKibben's 350.org global team. The U.N.'s 2,000 scientists know overpopulation is Earth's only real problem.

Get it? Earth has only one real problem, there's the one main dependent variable in the scientific equation. But we refuse to focus on it. So, yes, even scientists are science deniers too. They know population growth is the killer issue, but are avoiding it too. Thousands of scientists have brilliant technical solutions to reducing the impact of global warming. But avoid the root cause. They keep solving the dependent variables in their climate-change science equation. But population growth is the cause of the Earth's problem, not the result.

Stop, shift, focus on the real problem. Stop focusing on the dependent variables. Your scientific method makes this clear ... we are making too many babies. Population's out of control. And that's the world's No. 1 problem. But we're all in denial. So, nobody's dealing with the world's biggest problem. Listen:

- *Scientific American* says global population growth is "the most overlooked and essential strategy for achieving long-term balance with the environment." By 2050 world population will explode from 7 billion to 10 billion, with 1.4 billion each in India and China, and China's economy nearly three times America's.

- In "The Last Taboo," *Mother Jones* columnist Julia Whitty warned: "What unites the Vatican, lefties, conservatives and scientists in a conspiracy of silence? Population." But this hot-button issue ignites powerful reactions. So politicians won't touch it. Nor will U.N.'s world leaders. Even if it's killing us."

- Back in 2009 billionaire philanthropists meet secretly in Manhattan: Gates, Buffett, Rockefeller, Soros, Bloomberg, Turner, Oprah and others. Each took 15 minutes to present their favorite cause. Asked what was the "umbrella cause?" Answer: Overpopulation, said the billionaires. Too many people.

- Jeremy Grantham's investment firm GMO manages about $110 billion in assets. He also backs the Grantham Institute of Climate Change at London's Imperial College. He says population growth is a huge "threat to the long-term viability of our species when we reach a population level of 10 billion" because it is "impossible to feed the 10 billion people." We don't need more Big Ag farming, we need fewer mouths to feed.

- In *The Sixth Species Extinction,* Elizabeth Kolbert warns that "though it might be nice to imagine there once was a time when man lived in harmony with nature, it's not clear that he ever really did." Which increases the probability that as population increases and our limited resources get scarcer, wars will increase just as the Bush Pentagon warned, about the time the Iraq war was launched.

But how to tackle the problem? Bill Gates says let's cap global population at 8.3 billion, even as his vaccine and contraceptive plans extend life expectancy. Columbia University's Earth Institute Director Jeff Sachs says even 5 billion is unsustainable, too many people. To stop adding more is tough enough. But how do eliminate two billion from today's seven billion total already here? Voluntary? Remember China's one-child plan didn't work.

OUT-OF-CONTROL POPULATION HAS NO SOLUTION

Worst-case scenario: There is no solution. None. Out-of-control population is going to drive us off a cliff. Even worse, seems nobody really cares. Nobody's working on a real solution. No one has the courage. No political will. Not U.N. leaders, scientists or billionaires. No one. It's taboo. All part of a bizarre conspiracy of silence, and the denial is killing us.

Any real solutions? Or do we all just wait for wars, pandemics, starvation to erase billions? Wait in denial? Is that the sound of the sixth great species extinction dead ahead? Will pandemics, poverty, resource wars solve Earth's biggest problem, the problem no one wants to talk about—over-population?

Meanwhile, Big Oil's marketing studies keep telling CEOs like Tillerson the truth about the inconsistent behavior of irrational humans living in denial. How we just keep telling ourselves we're recyclers, green, love hybrids, eat organic.

Why? Because we just keep buying Fords, Jeeps and Teslas, keep holding Big Oil stocks for retirement, keep stocking up on carbon polluting products, because our subconscious secretly endorses Big Oil's strategy. As Tillerson told Charlie Rose: "My philosophy is to make money. If I can drill and make money, then that's what I want to do," making "quality investments for our shareholders."

CAN WE AVOID OUR OWN EXTINCTION? IS IT TOO LATE?

"One of the disturbing facts of history is that so many civilizations collapse," warns Jared Diamond, environmental anthropologist and author of the classic *Collapse: How Societies Choose to Fail or Succeed.* Many "civilizations share a sharp curve of decline. Indeed, a society's demise may begin only a decade or two after it reaches its peak population, wealth and power."

Can it be stopped? Before it's too late? Don't bet on it. Watching how Washington solves real problems lately is not encouraging. Diamond detailed the scenario that keeps repeating in history: We need leaders with "the courage to practice long-term thinking, make bold, courageous, anticipatory decisions at a time when problems have become perceptible but before they reach crisis proportions."

But unfortunately, leaders become rigid and myopic, driven more by personal interests than courage, long-term thinking and the public interest. They move from crisis to crisis, often too little, too late. And at some point, they pass the point of no return, are caught off-guard, make errors of judgment and their world's collapse, rapidly.

The dinosaurs didn't even know what hit them in the last great species extinction. But we do know what's ahead. Or do we? Are we just "sleepwalking to disaster," as the U.N. Secretary-General warns.

9.7.10

PEAK

FOOD

The "inevitable mismatch between finite resources and exponential population growth" is a "threat to the long-term viability of our species when we reach a population level of 10 billion," because it is "impossible to feed 10 billion people ... We have the brains and the means to reach real planetary sustainability. The problem is with us and our focus on short-term growth and profits, which is likely to cause suffering on a vast scale." Jeremy Grantham, billionaire philanthropist

PLANET-EARTH CANNOT FEED 10 BILLION PEOPLE

Forget Peak Oil. Refocus, on peak food. Yes, the planet's maxing out. We're maxing-out on Peak Food. Billions go hungry. We are poisoning our future. That's why Cargill, America's largest private food company, is warning us: about water, seeds, fertilizers, diseases, pesticides, droughts. You name it. Everything impacts the food supply. Wake up America, it's worse than you think, and soon too late.

Yes, we are slowly poisoning America's food supply, poisoning the whole world's food supply. Fortunately, Cargill is thinking ahead. Too bad our political leaders are dragging their feet. Trapped in denial, they blindly protect Big Oil donors, afraid of losing their job security. But their inaction is killing our future, starving, poisoning people, while they hide behind junk-science.

The honest-to-God truth is Big Ag's worldwide farm production growth cannot feed the 10 billion humans who will be living on Planet-Earth by 2050. We also can't wait till 2050 for the fallout. The clock's ticking on the "Peak Food" disaster dead ahead. We're at the critical tipping point, the planet is boiling over.

Conservative Greg Page, executive chairman of the Cargill food empire, has that great can-do spirit of capitalism: At $43 billion, Cargill is America's largest privately held company, launched during the Civil War with one grain warehouse. An unabashed optimist, Page was sounding a loud battle cry in Burt Helm's *New York Times* op-ed, "The Climate Bottom Line:"

Page is a powerful leader, optimistic, realistic, experienced ... admits he "doesn't know ... or particularly care ... whether human activity causes climate change ... doesn't give much serious thought to apocalyptic predictions of unbearably hot summers and endless storms." Yes, Page wants results ... for shareholders?

Page is obviously no left-wing environmentalist. Far from it. This is business, jobs, profits, because it's a fact, climate's already damaging huge sectors of America's agricultural business ... dust bowls in the heartland, in California's bone dry central valley, all over ... Georgia, North Carolina, Texas, all farm economics are affected. Meanwhile, our politicians dilly-dally, drifting, dragging their feet, in denial, playing petty ideological games.

For Cargill's boss, this really is "just business," because in the long run climate change could easily and irreversibly wreck Cargill's as well as America's nearly $10 trillion annual agricultural sector that employs over 2.5 million.

NEW AG TECHNOLOGIES EMERGE TO FEED 10 BILLION?

Yes, Page's Big Ag food company has billions at stake. He cannot risk being wrong, losing. Listen to the future he sees coming: "Over the next 50 years, if nothing is done ... crop yields in many states will most likely fall ... the costs of cooling chicken farms will rise ... and floods will more frequently swamp the railroads that transport food in the United States" ... he wants American agribusiness to be ready.

But what if Cargill's scientists are too optimistic, when arguing America's agriculture sector is "well prepared to adapt to changes." Maybe not. Former New York mayor, billionaire Michael Bloomberg, was skeptical of Cargill. The Times reported Bloomberg asking Page: "Do the technologies exist? Or are you saying they will someday, 'as in, we know there will be a cure for cancer, but we have no idea when or how'?"

Ouch! Predictably, Cargill, Big Ag, the entire global food industry will likely just keep focusing on short-term profits, living in their fantasy world, hoping that maybe someday—*like "energy miracle" Bill Gates hopes for*—maybe they'll get bailed out by some magical technology. Meanwhile, the clock's tick-ticking accelerates.

Bottom line: Cargill's Page is convinced "adaptation was more a matter of execution for the food industry, not research and development." Unfortunately, as things get worse and Peak Food starts descending rapidly down the other side of the peak, all that optimism, the magical technology, all the promises of capitalist solutions will still be little comfort for the more than one billion people worldwide who already live in extreme poverty.

SKEPTICISM POISONING STEEP 'PEAK HUNGER' DECLINE?

Yes, food is one of the biggest problems in the world: We already have trouble feeding the 7.3 billion people already on the planet today. And it's virtually impossible to feed another 3 billion by 2050, warns Jeremy Grantham, whose firm is an investment manager for $120 billion, also funds the Grantham Research Center at the London School of Economics.

Bill Gates maxes population out at 8 billion people. Columbia University's Jeff Sachs, head of the Earth Institute and a key adviser to the UN Secretary General, warns that 5 billion is the top that Earth can sustain. And still, at today's trajectory, we'll hit 10 billion by 2050 … a disaster is waiting to happen.

The Bush Pentagon already predicted that by 2020 the planet's "carrying capacity" will be so drastically compromised that America's war machine had already begun preparing military defense systems for the coming "all-out wars over food, water, and energy supplies."

Feed the 7 billion people already on Earth today, plus another 3 billion by 2050? Feed 10 billion? If we must, we can't wait till 2050 to start. The clock is ticking. Now. We're already near critical mass, the tipping point. We better start planning now, if we want even a fighting chance.

BIGGEST SURVIVAL TASK IS FOOD & AG TECH'S FAILING

Unfortunately, Grantham is not as optimistic as Page. Just the opposite, he reinforces the Pentagon's worst fears, warning of an "inevitable mismatch between finite resources and exponential population growth" with a "bubble-like explosion of prices for raw materials," plus commodity shortages that are a major "threat to the long-term viability of our species when we reach a population level of 10 billion," making "it impossible to feed the 10 billion people."

Bad news: the planet's "carrying capacity" cannot feed 10 billion people. So that constrains all technological optimism, forcing Grantham conclusion: "As the population continues to grow, we will be stressed by recurrent shortages of hydrocarbons, metals, water, and, especially, fertilizer. Our global agriculture, though, will clearly bear the greatest stresses."

Get it? Agriculture is the world's biggest problem for the commodity sector. Agribusiness has the "responsibility for feeding an extra two to three billion mouths, an increase of 30% to 40% in just 40 years. The availability of the highest quality land will almost certainly continue to shrink slowly, and the quality of typical arable soil will continue to slowly decline globally due to erosion despite increased efforts to prevent it, puts a huge burden on increasing productivity." An impossible equation for Cargill.

5 REASONS 'PEAK FOOD' IS WORLD'S TICKING TIME-BOMB

Grantham believes "humans have the brains and the means to reach real planetary sustainability." Instead "the problem is with us and our focus on short-term growth." Our "human ingenuity" can even solve the energy problem, even shortages of metals and fresh water. Even solve the population demand problem without starvation, diseases and wars.

But agriculture is facing a huge loss of nonrenewable resources that technology cannot solve with unsubstantiated promises of future magic. So, here's why agriculture is the world's No. 1 time bomb. And why American politicians damn well better start to deal with Grantham's five constraints:

1. We're completely running out of potassium (potash) and phosphorus (phosphates) and eroding our soils ... Their total or nearly total depletion ... by itself makes it impossible to feed 10 billion people.

2. Potassium and phosphorus are necessary for all life, they cannot be artificially manufactured and there are no natural substitutes.

3. Globally, soil is eroding at a rate that is several times that of the natural replacement rate.

4. Poor countries found mostly in Africa and Asia will almost certainly suffer from increasing malnutrition and starvation. The possibility of foreign assistance on the scale required seems remote.

5. Many stresses on agriculture will be exacerbated ...by increasing temperatures ... increased weather instability ... frequent and severe droughts, floods and wars.

Grantham is skeptical of solutions that are based on the usual short-term thinking will work in the future: "Capitalism, despite its magnificent virtues in the short term, above all, its ability to adjust to changing conditions, has several weaknesses." Capitalism "cannot deal with the tragedy of the commons, e.g., overfishing, collective soil erosion, and air contamination." Just the opposite, unregulated free markets just makes things worse.

And yet in today's culture of science denialism, the "finiteness of natural resources is simply ignored, and pricing is based entirely on short-term supply and demand." In short, the next few decades challenge a fundamental tenet of capitalism: That the public good is best served by the "invisible hand" of competing individuals, acting solely in their own separate special interests. No cooperation, no global solutions, it's everyone for themselves, no restrictions. Unfortunately, that's a dead-end for everyone, a time bomb ticking loudly.

2.11.15

DROWNING

INDEBT

"Peak Production From a Planet in Distress: Can We Keep It Up?"
Our global economic system is "programmed to squeeze ever more
resources from a planet in distress ... A mixture of population growth,
consumerism, greed, and short-term thinking by policy makers
and business people seems to be inexorably driving human
civilization toward a showdown with the planet's limits."
Earth Policy Institute and Worldwatch Institute

$60 TRILLION GAMBLE IN $75 TRILLION GDP WORLD

Fixing global warming is a huge $60 trillion gamble in a world where total debt is already about $225 trillion—more than 300% of the $75 trillion annual GDP of the world's 197 nations. In other words, the world's drowning in debt, near defacto bankruptcy, and global warming will make us sink deeper and faster.

And yet ... we just keep betting that somehow, it's not really happening ... going all-in at the high-stakes global casino ... blindly betting on capitalism's promise of perpetual growth ... that technology will save us ... addicts gambling away future generations ... the elite few risking everybody's future in order to increase their personal wealth today ... making an irreversible $60 trillion global warming bet where the odds of losing mean GDP collapsing below 1% by 2100.

Can this high-risk gamble be stopped? Maybe not. A recent *Scientific American* research study estimated cost to taxpayers at $60 trillion. Meanwhile, the costs, the risks, the fears just keep on growing bigger. *The Guardian* newspaper calls this the ticking "economic time bomb" that could "undermine the global financial system." Yes, kill economic growth worldwide, destroy America's future, end civilization as we know it. And still, our civilization keeps gambling away the future, trading everything for greater wealth today.

Political wars will accelerate: Big Oil, the GOP, a new president and all their climate science-deniers will double down. Now Ernst & Young researchers estimate "20% of the world's undiscovered but recoverable oil and natural gas resources" are under the methane-laden Arctic. So, our myopic Big Oil will fight even harder to protect their $100 billion annual profits, making certain not one dime of America's $17 trillion GDP is spent to stop the ticking time bomb.

It may already be too late said Bill McKibben in *Foreign Policy,* too late to stop the global warming time bomb. Big Oil keeps re-electing GOP climate-science deniers to Congress, a *ThinkProgress* headline read: "Steve King: Belief in Climate Change is a 'Religion, Not Science'." An Iowa Republican, he spoke at "an event sponsored by the climate-denying, Koch-funded Americans for Prosperity."

DENIALISM TRIGGERING CATASTROPHIC LOSSES

"Sometimes facing up to the truth is just too hard. When the facts are distressing it is easier to reframe or ignore them," writes Australian Professor of Ethics Clive Hamilton in his classic, *Requiem for a Species: Why We Resist the Truth about Climate Change*: "Around the world only a few have truly faced up to the facts about global warming ... It's the same with our own deaths; we all 'accept' that we will die, but it is only when our death is imminent that we confront the true meaning of our mortality."

But soon the impact of global warming will be too obvious, costly, disastrous to ignore, even for the Koch Bros and other Big Oil capitalists. When reality hits their bottom line, they will change. But by then, it will be too late for America, Miami and New York, Nantucket, Malibu, Silicon Valley, and all across the world, in Africa, Australia, China, South America, Oceana because the momentum can't stop, just keeps us drowning.

<u>AMERICA</u>: COASTAL CITIES FALLING INTO THE OCEANS

After Superstorm Sandy, *BusinessWeek's* cover was clear, "It's Global Warming, Stupid." But America quickly went back to sleep. Then, in *Vanity Fair's* "From Coast to Toast" we learned that "two of America's most golden coastal enclaves are waging the same desperate battle against erosion. With beaches and bluffs in both Malibu and Nantucket disappearing into the ocean, wealthy homeowners are prepared to do almost anything—*spend tens of millions on new sand, berms, retaining walls, and other measures*—to save their precious waterfront properties." This is a war between the rich and the rest, "deep-pocketed summer people and local working folks."

Then in "Goodbye Miami" *Rolling Stone* magazine covered more water problems: "By century's end, rising sea levels will turn the nation's urban fantasyland into an American Atlantis. But long before the city is completely underwater, chaos will begin." Other cities at risk: New York. San Francisco, New Orleans, Boston, Norfolk, Galveston, Tampa, even Silicon Valley which sits "by the edge of the bay protected by old levees that could easily fail."

CHINA: CANCEROUS WATER YOU CAN'T SWIM IN OR DARE DRINK

Time magazine: "After more than three decades of economic prosperity, China faces serious environmental challenges ... filthy waterways." After his sister died of cancer, Jin Zengmin, a rich entrepreneur, offered "a $32,000 reward to the chief of the local environmental-protection department if he dared to swim in a nearby river for a mere 20 minutes." He refused to swim in the garbage-infested water. "The river is poisonous," said Jin, "if we Chinese die of cancer caused by pollution, what's the meaning of economic growth for us?

AFRICA: WATER LOSS KILLING FUTURE ECONOMIC GROWTH

BusinessWeek warns: "Water, or the lack of it, is one of the biggest issues facing urban Africa, which will see a 66% population increase to 1.2 billion people by 2050 ... Although water shortages have long plagued parts of the continent, they've become the potential killer of Africa's economic takeoff. Ghana's $35 billion economy, whose estimated growth of 8% in 2013 would outpace the sub-Saharan African average for a sixth straight year, cannot continue at that rate without a modern water network."

SOUTH AMERICA: CAPITALIST PROFITS VS DRINKING WATER

Bloomberg Markets magazine warns of "Deadly Water Wars" where "governments and big, often foreign-based companies across Latin America are battling over water with families, communities and farms." Roots of their water wars: "Leaders across the region, elected on promises to fuel economic growth and lift populations out of poverty, are fast tracking water-use approvals for mining and other industrial uses. ... Water isn't always where the best mineral or agricultural riches are located so people are losing homes and farms as water is diverted to industrial uses." And the inequality gap keeps widening.

AUSTRALIA: WATER CRISIS 'DISASTERS OF BIBLICAL PROPORTIONS'

In another *Rolling Stone* feature, "The End of Australia," Jeff Goddell asks: "Want to know what global warming has in store for us? Just go to Australia, where rivers are drying up, reefs are dying, and fires and floods are ravaging the continent." University of Melbourne climate researcher David Karoly warns: "Australia is the canary in the coal mine ... What is happening in Australia now is similar to what we can expect to see other places in the future."

INDIAN OCEAN: VILLAGES FLOOD, ISLANDS SINK, POWER POLITICS

The *New York Times* reviewed a documentary, "The Island President." Mohamed Nasheed is a former president of Maldives, a nation of 1,200 islands in the Indian Ocean. Nasheed warned that "rising sea levels could wipe out his country." Violence erupted: "Protesters burned police stations in the south, and Islamic radicals smashed nearly 30 Buddhist statues dating to the sixth century in the National Museum." Nasheed spoke at the 2009 UN Climate Change Conference in Copenhagen. He was "forced out because he threatened the interests of the old order that ran the country for 30 years."

CENTRAL AMERICA: MAYAN CIVILIZATION, WARNING FOR FUTURE

The *World Wildlife Foundation* interviewed Jared Diamond, author of *Collapse: How Societies Choose to Fail of Succeed.* Diamond explained the lessons of climate change learned from the centuries of a Mayan civilization of two million people: "There are so many societies in which the elite made decisions that were good for themselves in the short run and ruined themselves and societies in the long run ... The Mayans collapsed "because of a combination of climate change, drought, water-management problems, soil erosion, deforestation ... The kings had managed to insulate themselves from the consequences of their actions."

And yes, they saw forests "being chopped down." But "kings didn't recognize that they were making a mess until it was too late, when the commoners rose in revolt. ... Similarly, in the United States at present, the policies being pursued by too many wealthy people and decision-makers are ones that—as in the case of the Mayan kings—preserve their interests in the short run but are disastrous in the long run."

SOUTHEAST ASIA: ANKOR, WATER DESTROYED GREAT CIVILIZATION

National Geographic featured the ancient empire: "Greater Angkor probably encompassed between 600,000 and 1 million inhabitants, at a time when London had perhaps 30,000 people." Built on the edge of Southeast Asia's "Great Lake" the people needed a vast irrigation system: "During the monsoon season, vast amounts of water poured through the watershed causing the Mekong River to actually back up behind its delta."

For centuries the people relied on these vast waterworks for crops, fish, drinking. But "the very system that allowed the Khmer to support such a large population may have been their undoing." Slowly: In the mid-1200s, a flood "destroyed part of the earthworks." A century later "monsoons became very unpredictable. ... An extreme drought crippled what remained of the once-glorious Khmer Empire, leaving it vulnerable to repeated attacks and sackings by the Thais. By 1431 ... power shifted south" to coastal cities.

NEW 'DENIAL OF DEATH:' BIG OIL, CONSUMERISM, GDP

One final observation on the world's new love of free-market capitalism and its blind obsession with economic growth at all costs: *BusinessWeek* quoted David Owen from his *The Conundrum:* "As long as the West places high and unquestioning value on economic growth and consumer gratification, with China and the rest of the developing world right behind, we will continue to burn the fossil fuels whose emissions trap heat in the atmosphere," driving us way past the point of no return.

Yes, catastrophes will shock us along the way, all across the world, and soon, jolt us out of our denial. But it'll be too late to act. Hamilton explains our denial trap: "Around the world only a few have truly faced up to the facts about global warming ... It's the same with our own deaths; we all 'accept' that we will die, but it is only when our death is imminent that we confront the true meaning of our mortality."
9.29.14

BUY IN$URERS!

SELL BIG OIL, FOSSIL FUELS

*"New scientific evidence that the world's oceans
warmed significantly. Ocean energy is the primary cause
of extreme climate events, increasing the number of
insurance-relevant hazards, a near irreversible shift.
Even if greenhouse gas emissions stopped,
ocean temperatures would keep rising."*
Geneva Association: Study of Insurance Economics

NEW MEGATRENDS TRANSFORM INSURANCE RISK ANALYSIS

Not long ago, insurers operated on simple rules-of-thumb—*like 100-year mega-fires, 100-year mega-floods, 100-year droughts, and all kinds of "100-year" rules-of-thumb*—to explain climate disasters that happened infrequently, often generations apart. Rules-of-thumb that seemed to work. Or so they thought.

But today, it seems 100-year climate disasters have begun hitting the news media every 100 days. And the frequency continues accelerating: Meteorologists are now reporting powerful once-in-a-1,000-year "biblical" disasters increasing in frequency as costs keep skyrocketing.

USA Today headlines capture the accelerating trends: "1,000-Year Storm Slams S.C." Then the next day it was: "Biblical Flooding Becoming More Common." Yes, climate disasters are now more frequent, more costly and as a result better long-term investments than Big Oil and fossil fuels.

Next this report; "Biblical flooding in South Carolina is at least the sixth so-called 1-in-1,000 year rain event in the U.S. since 2010, a trend that may be linked to factors ranging from the natural, such as a strong El Niño, to the manmade, namely climate change," according to the National Oceanic and Atmospheric Administration (NOAA).

SIX "ONCE-IN-1,000-YR BIBLICAL" DISASTERS IN SIX YEARS

Worse, looking back over breaking news from the past decade tells an even bigger global story of disasters ... remember the Arizona megafires, killed 19 firefighters ... the recent megafires sweeping Oregon, Washington ... Oklahoma tornadoes ... earlier Jersey Shore's Superstorm Sandy ... the Yosemite Rim Fire ... out of control Colorado flooding... hurricanes in Mexico ... Africa floods ... Japan's typhoons ... India's monsoons ... China's earthquakes ... California Central Valley farm economy ground water disappearing.

Climate-disaster news is relentless, increasing, we tune it out to keep our sanity. Disaster after disaster, costing billions on top of billions. Yes, our world is being hit with an accelerating barrage of bigger, more powerful, costlier natural disasters. Yet in a parallel fantasy universe Big Oil, conservative billionaires and their GOP lobbyists remain with their heads in the sands, trapped in their predictable irrational denial of the increasing negative impact of climate change.

Worldwide, the insurance industry is waking up the public, creating a new paradigm shift. Until now, climate-science-denying capitalists, Big Oil, Koch billionaires and the entire fossil-fuel industry all assumed the American taxpayer will just keep picking up the tab forever ... that climate change is not manmade ... assumed Americans are the insurer-of-last-resort ... that climate disasters are really 'Acts of God'—His fault—so the public had to foot the bill, self-insure.

Warning, folks, that's all changing: Insurers are capitalists. They're in business to make a profit, same as Exxon and the Koch Bros. Insurers are no longer relying on 19th century rules of thumb and outdated formulas. Nor can they let the myopic rhetoric of today's science deniers, whether Congress, fossil fuel giants or GOP state governors set insurance rates and policies. Things are changing with new insurance industry players on the offensive.

NEW SCIENCE VS OLD "100-YEAR DISASTER" RULES-OF-THUMB

With today's new high-tech world, science-driven, big-data capitalism rules, insurers can no longer make money using outdated, seat-of-the-pants, hand-me-down 100-year and 1,000-year disaster reoccurrence formulas that no longer reflect the reality of our 21st century's accelerating trends where everything's happening at hyper speed, tweets, trades, travel and now climate traumas. For we now know, so-called "100-year disasters" are too often hitting every 100 days, and so-called "1,000-year biblical disasters" can repeat annually.

So, climate science deniers can babble on all they want—Big Oil, Koch Bros, Chamber of Commerce, and their GOP lobbyists—can rant and rave all they want about their myopic ideologies, unprincipled obstructionism, underlying greed, and tout their so-called anti-scientific research. They can pay off academicians to write articles and make speeches to cast doubt on legitimate climate science. And they can buy all the national ads they want to degrade climate activists like Michael Bloomberg, Hank Paulson, Tom Steyer and their RiskyBusiness.org, Greenpeace and Bill McKibben's 350.org global army of environmental activists.

But soon, all the science-deniers' noise may not matter much. Profit-minded insurance industry capitalists are wising up. They know the Age of Big Oil has peaked, the fossil fuel industry is dying. So new climate warming legislation may even be needed ... no grand bipartisan political bargain may be necessary ... nor any long, drawn-out, costly lobbying efforts called for ... no new worldwide campaigns to support alternative energy innovations ... indeed, even the UN Paris Agreement will be irrelevant on a dying planet.

Why? In the future, the increased frequency, intensity will be so obvious, the scientific data so overwhelming, that insurance costs will finally have to be based not on mythic rules-of-thumb but on hard evidence, climate-science statistics pushed are the real culprits.

SIX GLOBAL MEGATRENDS CHANGING INSURANCE INDUSTRY

The Geneva Association for the Study of Insurance Economics, a global think tank, published a report that pinpoints a new direction for climate-change insurers: "Warming of the Oceans And Implications for the (Re)insurance Industry." Formed in the 1970s, the Geneva Association is now the industry's leading global insurance think tank for "strategically important insurance and risk-management issues."

The Geneva Association is a recognized spokesman for the world's most powerful insurance groups. Membership is made up of 90 CEOs from the world's top insurance and reinsurance companies, operating worldwide through a network of industry power players. They develop strategies for risk management in today's uncertain global economy. Their "annual General Assembly is the most prestigious gathering of leading insurance CEOs worldwide."

Recently the association identified the world's six biggest trends that are having the greatest impact on the future of climate change and insurers. They warn that climate and global-warming problems are bad, getting worse. Keep in mind that the Geneva Association for the Study of Insurance Economics is not an organization of left-wing activists but hard-nosed profit-motivated capitalists. And their six megatrends are based on the same hard scientific evidence that climate-science deniers are categorically dismissing. Here's a summary of the Geneva Association report on the six megatrends:

1. GLOBAL WARMING ACCELERATING, SOON IRREVERSIBLE

"New scientific evidence that the world's oceans ... warmed significantly ... ocean energy is the primary cause of extreme climate events ... increasing the number of insurance-relevant hazards ... a near irreversible shift ... even if greenhouse gas emissions stopped, ocean temperatures would keep rising."

2. RISK ANALYSIS, SCIENTIFIC DATA REPLACING OLD RULES

Geneva's report is quite clear, "traditional approaches based solely on analyzing historical data no longer work in making risk assessments today ... A paradigm shift from historic to predictive risk assessment methods is necessary" using new "scenario-based approaches and tail risk modeling." Science facts are replacing climate-denier capitalist ideologies.

3. CLIMATE CHANGE HAVING HUGE IMPACT ON WORLD ECONOMY

Warning, "in some high-risk areas, ocean warming, and climate change threaten the insurability of catastrophic risk more generally." A separate *Scientific American* research study estimates the cost to global economies at $60 trillion. *The Guardian* newspaper of London called it an "economic time bomb" that will "undermine the global financial system" of a world whose total GDP is only $75 trillion.

4. WEATHER EXTREMES: "DRIER DRIES! ... WETTER WETS!"

"Warmer oceans ... means more water ... warmer atmosphere ... more energy ... increasing the intensity of extreme events ... associated precipitation ... and increasing the loss potential of natural catastrophes."

5. RISING SEA LEVELS FAST REDUCING ECONOMIC GROWTH

"Thermal expansion of the oceans ... melting of continental ice shelves and glaciers has increased global sea levels ... the rate is accelerating ... rising sea levels increase the risk of flooding ... storm surges ... decreasing the protective life span of coastal infrastructures such as Dutch flood dikes or the Thames barrier ... Sea level rise also increases the damage potential from geophysical events."

6. UNPREDICTABLE MEGATRENDS INCREASING RISK COSTS

"Warming of the oceans ... affecting ... large-scale climate patterns ... However, due to the long time scales of ocean dynamics ... and the relatively short length of observational data ... the effects of those changes on catastrophic risk ... are unclear."

Bottom line on climate change: McKibben wrote in *Foreign Policy* that it "might already be too late." Today, the smart money is betting on the re-insurance industry's shifting paradigm: For too long Big Oil and climate-science deniers have been shifting the burden of their short-term profit strategies to the public. But the new scientific data from the re-insurance industry is changing all that. The plus side reveals new opportunities for investors—buy insurers, sell fossil fuels.

10.12.15

REVERSING

EVOLUTION

"Over the last several decades, human activities have so altered the basic chemistry of the seas that they are now experiencing evolution in reverse: a return to the barren primeval waters of hundreds of millions of years ago." Alan Sielen, Scripps Institute of Oceanography

10 WAYS CAPITALISM'S REVERSING PLANET'S EVOLUTION

Yes, across the world capitalists are getting rich off the high seas, a vast reservoir of wealth holding 95% of the planet's water, spanning 70% of the Earth's surface. Often called the "last frontier," a return to America's 18th century Wild West, they are virtually unregulated, a new free market where capitalists roam like pirates, plundering wealth and treating our oceans as a freebie gold mine and trash dump.

Bad news for seven billion people living on the planet. And sometime around 2050 we'll be adding three billion more people. Scientists already know the planet can't feed 10 billion. Now we're polluting their water. Won't be enough clean water for all to drink, triggering resource wars.

The bad news gets worse: in "The Devolution of the Seas. The Consequences of Oceanic Destruction," a challenging piece in *Foreign Affairs* journal, Alan Sielen of the Scripps Institute of Oceanography warns: "Over the last several decades, human activities have so altered the basic chemistry of the seas that they are now experiencing evolution in reverse: a return to the barren primeval waters of hundreds of millions of years ago."

TURNING BACK 'EVOLUTIONARY CLOCK' BILLION YEARS

Evolution in reverse? Yes, Planet-Earth is regressing eons to an earlier primitive era. Unregulated free-market competition on the high seas is turning back our evolutionary clock. That doesn't bother today's short-term-thinking capitalists. But it should. Because, ironically, shifting evolution into reverse will also self-destruct the very global economy that capitalism needs for future growth.

Yes, today's capitalists see the next three billion people born on the planet only as the new customers essential to the expansion of free market capitalist globally. They ignore the reality that nature has designed all species with built-in termination programs. Deny it all you want, but every biological species, like every individual, has entrances and exits, as Shakespeare put it. Eventually we must say goodbye.

Capitalists will deny the dark impact they have on our global economy, dismissing the long-term economic reality of treating the world's oceans as a free garbage dump. Capitalism is programmed to focus on the short term, on profits, high frequency trading, in microseconds, on day-end closing prices, quarterly earnings, annual bonuses. Rarely longer. They ignore the self-destruct button that will eventually boomerang, trigger the endgame for capitalism, their society, the human race, our planet's water resources.

Capitalists are living in a new Wild West World. No lawmen, just free-market competitors, free to do whatever they want, whenever, unregulated, uncontrolled, unrestrained, skimming, mining, plundering the wealth of the high seas, free to use, misuse and abuse vast oceans of water—and at no cost to their profits. Here's how the Scripps Institute of Oceanography's Sielen summarizes the issues: 10 costly ways that capitalists increase their short-term profits, leaving the long-term losses to the masses worldwide, for the next generation to pay:

1. POLLUTERS TREAT OCEANS AS FREE GARBAGE DUMP

Sielen's imagery is powerful: "The oceans' problems start with pollution, the most visible forms of which are the catastrophic spills from offshore oil and gas drilling or from tanker accidents." But that "pollution pales in comparison to the much less spectacular waste that finds its way to the seas through rivers, pipes, runoff, and the air ... trash, plastic bags, bottles" washing into "coastal waters or discarded by ships ... drifts out to sea ... forms epic gyres of floating waste" covering "hundreds of miles."

2. DESTRUCTION OF MARINE LIFE: LOST JOBS, FOOD, ETC

"The prospect of vanishing whales, polar bears, Bluefin tuna, sea turtles, and wild coasts should be worrying enough on its own," warns Sielen. "But the disruption of entire ecosystems threatens our very survival, since it is the healthy functioning of these diverse systems that sustains life on earth." Destruction on this level" has massive consequences, and costs "humans dearly in terms of food, jobs, health, and quality of life."

3. TOXIC CHEMICALS POLLUTING WORLD'S OCEANS

It gets far worse: "The most dangerous pollutants are chemicals" poisoning the oceans with toxins, says Sielen. They "travel great distances, accumulate in marine life, and move up the food chain." Mercury from burning coal "rains down on the oceans, rivers, and lakes." Each year hundreds of new untested industrial chemicals "build up slowly in the tissues of fish and shellfish," get passed to larger creatures and humans "causing death, disease, and abnormalities," adversely affecting "development of the brain, the neurologic system, and the reproductive system in humans."

4. FERTILIZERS DEPLETING OXYGEN, CREATING DEAD ZONES

Add to this disaster scenario, fertilizers, which are now being added in excessive levels causing "havoc on the natural environment," an explosive growth of algae, decomposition, a loss of "oxygen needed to support complex marine life," creating "dead zones devoid of the ocean life" that "have more than quadrupled" in a decade.

5. HUMANS EATING TOO MANY FISH, OTHER SEA FOODS

World population was less than 3 billion in 1950, 6 billion in 2000. Projected at 10 billion in 2050. Sielen warns: "Humans are simply killing and eating too many fish," with fish supplies falling "dramatically." Tuna, swordfish, halibut, flounder populations dropped 90% since 1950. "Human appetite has nearly wiped those populations out."

6. FISH SUPPLIES DISAPPEAR, DEMAND KEEPS INCREASING

And demand keeps growing as supplies "are rapidly dwindling." Exploding prices add to the demand: This year, a 489-pound Bluefin tuna sold for $1.7 million making it "profitable to employ airplanes and helicopters to scan the ocean for the fish that remain; against such technologies, marine animals don't stand a chance." Small fish like sardines, anchovy, herring, are also disappearing, meaning less food for bigger fish up the food chain.

7. DESTRUCTIVE & WASTEFUL FISHING METHODS

Unfortunately, "modern industrial fishing fleets drag lines with thousands of hooks miles behind a vessel," with "nets thousands of feet below the sea's surface." Untargeted seals, turtles, dolphins, whales, albatross get entangled, killing millions of tons each year. "Some of the most destructive fisheries discard 80% to 90%." In the Gulf of Mexico, "for every pound of shrimp ... over three pounds of marine life is thrown away."

8. DESTROYING MARINE HABITATS KILLS FUTURE GROWTH

Another factor destroying our oceans: "The destruction of the habitats that have allowed spectacular marine life to thrive for millennia," says Sielen. And yet capitalism continues the "wholesale destruction of deep-ocean habitats ... submerged mountain chains called seamounts," some higher than Mt Rainier. They are "homes to a rich variety of marine life." Yet, industrial trawlers bulldoze their way" destroying "deep cold-water corals, some older than the California redwoods."

9. ACID BUILDUP IN OCEANS WEAKENS MARINE LIFE

"The buildup of acid in ocean waters," warns Sielen "reduces the availability of calcium carbonate, a key building block for the skeletons and shells of corals, plankton, shellfish, and many other marine organisms" that need it "to grow and also to guard against predators."

10. OUR HOME PLANET IS IN HOT, HOT WATER!

Echoing findings by 2,000 scientists in the recent Fifth Intergovernmental Panel on Climate Change, Sielen says scientists predict "climate change will drive the planet's temperature up by between 4 and 7 degrees Fahrenheit over the course of this century" causing "hotter oceans ... rising sea levels ... stronger storms" as the "life cycles of plants and animals are upended, changing migration patterns, causing other serious disruptions ... surface waters mixing less with cooler, deeper waters," reducing phytoplankton population, "the foundation of the ocean's food chain." For more than two decades U.N. climate agreement negotiations have exposed how capitalism's power-players control governmental decision-makers, prompting Sielen's final warning: "So long as pollution, overfishing, and ocean acidification remain concerns only for scientists ... little will change ... Diplomats and national security experts, who understand the potential for conflict in an overheated world, should realize that climate change might soon become a matter of war and peace ... Business leaders should understand better than most the direct links between healthy seas and healthy economies ... And government officials, who are entrusted with the public's well-being, must surely see the importance of clean air, land, and water."

"The world faces a choice," warns Sielen: "We do not have to return to an oceanic Stone Age," but that's where we're headed. To avoid that from happening we need to "summon the political will and moral courage to restore the seas to health before it is too late." The risks are enormous ... the odds very long ... the denial overwhelming ... the fuse burning short ... and the consequences not only catastrophic but irreversible.
12.4.13

GOLDEN CALF
OF CAPITALISM

*"Inequality is the root of social ills. As long as the problems
of the poor are not radically resolved by rejecting the
absolute autonomy of markets and financial speculation
and by attacking the structural causes of inequality,
no solution will be found for the world's problems or,
for that matter, to any problems." Pope Francis*

DESTROY GOD'S CREATION? IT WILL DESTROY YOU!

Fortune magazine celebrated the anniversary of anti-capitalist, socialist Pope Francis by ranking him first among the "World's 50 Greatest Leaders." Today the pope's in an ideological war with Big Oil backed GOP as well as conservative leaders everywhere. Yes, at war in America's unstable, endlessly fickle, political arena, a very long war, with pollsters warning that partisan politics may be pushing aside the pope's pronouncements.

Big picture: Economics leads the political soap opera. Lurking in the shadows, a new crash. Inevitable. And like 2000, nobody will hear it coming ... hidden under irrational exuberance, dot-com mania, millennium celebrations ... followed by a 30-month bear recession ... later the 2008 crash ... Alan Greenspan, Henry Paulson clueless ... already two crashes this century ... $10 trillion losses each ... next one coming after 2016 election cycle, with echoes of the McCain loss ... yes, a bigger badder bear than 2000 and 2008 ... because once again bulls and optimists, traders and leaders fall into denialism ... blinded by a raging new irrational exuberance.

What about the promise of big political changes? The GOP Congressional leadership talks a good game, but their anxious, conservative GOP base is sitting on the shifting tar sands of Big Oil cash threatened by higher costs, long-term risks. Yes, talk is cheap, but once partisan conflicts blow up, climate disasters will bury the GOP's aggressive energy agenda, support will fade.

Yes, Pope Francis has laid down his anti-capitalism manifesto for his army of 1.2 billion Catholics worldwide. He has also been removing conservative cardinals and bishops from leadership roles. He's hell-bent on changing the world fast. And his mandate is unwavering and unequivocal. He's drawing clear moral and political battle lines against repressive capitalism, excessive consumerism, rigid conservatism. Listen:

"Inequality is the root of social ills ... as long as the problems of the poor are not radically resolved by rejecting the absolute autonomy of markets and financial speculation and by attacking the structural causes of inequality, no solution will be found for the world's problems or, for that matter, to any problems." Yes, it sure sounds like a declaration of war: The anti-capitalist Pope Francis versus America's self-destructive amoral capitalism.

HOLY WAR: POPE'S MORALITY VS CAPITALIST ATHEISM

Pope Francis's target is clear: economic inequality is the world's No. 1 problem. Capitalism is at the center of all problems of inequality. And he speaks with a powerful moral authority — something missing from most American political leaders who are ideologically guided by atheist Ayn Rand, patron saint of the GOP's capitalism agenda in this moral war.

Without moral grounding, the GOP is no match for Francis' vision, his principled mandate, his long-game strategy to raise the world's billions out of poverty, to eliminate inequality, to attack the myopic capitalism driving today's economy, markets, politics.

Moreover, the pope has the resources: As commander-in-chief of the world's largest army: 1.2 billion Catholics worldwide who are now motivated to defeat capitalism's grip on inequality. His army includes 78 million Americans in 17,645 parishes, plus a huge officer corps of 213 cardinals, over 5,000 bishops, 450,000 priests and deacons worldwide, all sworn to carry out his vision. He needs no legislative approvals; popes have authority to act unilaterally, with speed, a dictator whose word is law, commanding allegiance, obedience and action.

THE POPE'S NEW 10 "ECONOMIC COMMANDMENTS"

So welcome to his first-anniversary celebration: Yes, it was just one year ago Pope Francis laid down his anti-capitalism agenda as a battle plan for Catholics. Mitch McConnell probably hasn't even read it, but every American should. So, here's an edited copy of Francis' 10 economic commandments. They define the specific strategies guiding his economic war against inequality and capitalism.

Here are Pope Francis's 10 strategies, in his own words directly from his "Apostolic Exhortation," a 67-page manifesto published by the Vatican before Thanksgiving last year. Listen to how driven Pope Francis is to changing the world. He really is a radical anti-capitalist, socialist, revolutionary leader with a message that American conservatives and capitalists worldwide, will face for years, long after any one politician's tenure. Listen:

1. SOLVE INEQUALITY BEFORE CAPITALISM DOOMS WORLD
"Inequality is the root of social ills ... as long as the problems of the poor are not radically resolved by rejecting the absolute autonomy of markets and financial speculation and by attacking the structural causes of inequality, no solution will be found for the world's problems or, for that matter, to any problems."

2. CAN'T TRUST "INVISIBLE HAND" OF FREE-MARKET GREED

"We can no longer trust in the unseen forces and the 'invisible hand' of the market. Growth in justice requires more than economic growth ... a better distribution of income ... The economy can no longer turn to remedies ... such as attempting to increase profits by reducing the work force and thereby adding to the ranks of the excluded."

3. TRICKLE-DOWN ECONOMIC IDEOLOGY OF RICH IS A HOAX

Some "continue to defend trickle-down theories which assume that economic growth, encouraged by a free market, will inevitably succeed in bringing about greater justice and inclusiveness in the world. ... a crude and naïve trust in the goodness of those wielding economic power ... the culture of prosperity deadens us."

4. TYRANNY OF CAPITALISM: THE RICH STEAL FROM THE POOR

"While the earnings of a minority are growing exponentially, so too is the gap separating the majority from the prosperity enjoyed by those happy few. This imbalance is the result of ideologies which defend the absolute autonomy of the marketplace and financial speculation, and reject the right of states charged with vigilance for the common good. ... A new tyranny ... unilaterally and relentlessly imposes its own laws and rules."

5. WORSHIP OF MONEY IS NEW "GOLDEN CALF" IDOLATRY

"Money must serve, not rule ... The current financial crisis can make us overlook the fact that it originated in a profound human crisis: the denial of the primacy of the human person! ... The worship of the ancient golden calf has returned in a new and ruthless guise in the idolatry of money ... lacking a truly human purpose."

6. CAPITALISM IS FUELING EXCESSIVE CONSUMERISM

"Today's economic mechanisms promote inordinate consumption, yet it is evident that unbridled consumerism combined with inequality proves doubly damaging to the social fabric. ... Inequality eventually engenders a violence ... new and more serious conflicts. Some ... blaming the poor and the poorer countries themselves for their troubles ... more exasperating ... widespread and deeply rooted corruption found in many countries ... businesses ... institutions."

7. COMPETITION FOR MORE WEALTH IS KILLING DEMOCRACY

"Today everything comes under the laws of competition and the survival of the fittest, where the powerful feed upon the powerless ... masses of people find themselves excluded and marginalized: without work, without possibilities, without any means of escape. ... Such an economy kills ... it is not a news item when an elderly homeless person dies of exposure, but it is news when the stock market loses two points?"

8. HUMANS AS "LEFTOVERS IN A THROWAWAY ECONOMY"

"Human beings are themselves considered consumer goods to be used and then discarded. We have created a 'throw away' culture which is now spreading. It is no longer simply about exploitation and oppression, but something new. ... those excluded are no longer society's underside ... no longer even a part of it. ... but the outcasts, 'leftovers'."

9. EXTREME INDIVIDUALISM SABOTAGING DEMOCRACY

"In a culture where each person wants to be bearer of his or her own subjective truth, it becomes difficult for citizens to devise a common plan which transcends individual gain and personal ambitions. ... freed from those unworthy chains and to attain a way of living and thinking which is more humane, noble and fruitful, and which will bring dignity to their presence on this earth."

10. CAPITALISM REJECTS GOD, MORALITY, BREEDS ANARCHY

"Behind this attitude lurks a rejection of ethics and a rejection of God. ... condemns the manipulation and debasement of the person. ... Ethics leads to a God who calls for a committed response which is outside the categories of the marketplace and makes it possible to bring about balance and a more humane social order."

We cannot afford to stand by passively. Join the dialogue, tell your friends about Pope Francis's radical plan to save the world from itself. Inequality impacts everyone. You, your kids, grandkids, the world, everyone. Nobody's on the sidelines in this new "Holy War."' Pass this on. Post the 10 new economic commandments on social media, tweet, retweet, let others know where you stand. The pope's 10 new economic commandments can save the world, save capitalists, save conservatives, save democracy. Yes, hope does spring eternal.

11.18.14

QUESTION #4:

"THE SOLUTION TO LIVING IN PEACE AND PROSPERITY ... ON A DYING PLANET?"

NEW EARTH

CONSCIOUSNESS

"The New Earth arises as more and more people discover their main purpose in life is to bring the light of consciousness into this world and so use whatever they do as a vehicle for consciousness." Eckhart Tolle, A New Earth: Awakening to Your Life's Purpose

---

LIVING IN PEACE & PROSPERITY ... ON A DYING PLANET!

Yes, our home planet is dying. Good news? You bet. *The Guardian* headline: "Warming of oceans due to climate change is unstoppable." Yes, they said *"climate change is unstoppable!"* One of more than two hundred climate news stories I get weekly from around the world. Planet-Earth is on a fast track to burnout. Ending up like Mars, the Red Planet? Good news? Yes. Listen:

Of course, climate science deniers like Big Oil billionaires and Senator Jim Inhofe dismiss all this as liberal activist hype. Good old American ingenuity and Silicon Valley tech geniuses can solve anything, right? Save the world. And for America's liberal left activists, progressives and environmentalists, hope springs eternal, there's always a bright light at the end of the tunnel.

Warning: Our planet really is dying, that's America's New Normal. Here's why:

Even though the UN Intergovernmental Panel on Climate Change gets all 200 member nations signed onto a climate accord in 2015, with the support of Pope Francis and President Obama, they freeze greenhouse gas emissions, climate scientists still predict "the sea will actually continue to warm for centuries and millennia ... and sea levels will continue to rise."

How much will sea levels rise? James Hansen, former director of NASA's Goddard Institute of Space Studies, warned Congress about all this 25 years ago. He just updated this study. It's brutally honest. There's a disaster dead ahead, a 10-foot sea level rise in 50 years, even higher in later decades. Yes, 10 feet. New Superstorm Sandys. Manhattan, Miami, underwater.

The United Nations Environmental Report's annual update went further. In an *InsideClimateNews* report, "Last Chance to Limit Global Warming to Safe Levels, UN Scientists Warn," John Cushman said: *"The next three years provide the last chance to limit global warming to safe limits in this century."* The U.N. says that "unless nations move before 2020 to cut their emissions more aggressively than they have promised, the window of opportunity will close." Apparently, our planet is not only dying it's the endgame is accelerating.

THE CONSCIOUSNESS OF THE NEW MASTER ALGORITHM

Optimism made America the greatest nation in history. The American Dream, our can-do spirit, our sense of being the "chosen," exceptional, challenged, winners, like our mission to the Moon. We build the future on hope and renewal. Even as we admit future trends are already baked in, unstoppable, inevitable. Admit minds are made up, rusted rigid.

Forget the relentless warnings made by the 2,500 scientists working on UN-IPCC research since 1988. They changed few minds. Our stone-deaf fossil-fuel conspiracy of deniers fear efforts to tax and regulate carbon emissions. Even as they still have billions in war chests to defeat scientific claims, pay off privately fund academic deniers, buy senators' loyalty, and then lie about their hypocrisy for decades.

Yes, our planet's dying, that's our civilization's new normal. Live with it. Forget about exposing the elected officials like Texas Sen. Ted Cruz (scientists are "cooking the books"), Sen. Jim Inhofe (global warming is the "Greatest Hoax"), and Sen. Mitch McConnell (protecting Kentucky's dying coal economy). Just three of the three hundred GOP favorites who get tens of millions in campaign cash to protect the fossil fuel industry.

Same goes with the Koch Bros billionaires. Their mindset is rigid. They have ten times more money than Trump, far more power and influence, already spending $1 billion to push the 2016 election further right for conservative causes, like killing EPA power plant regulations. They are of the New Normal.

And forget about repeating passages from Pope Francis's Climate Encyclicals. Yes, he's got a lock on the moral high ground and an army of billions worldwide. But campaign cash buys votes faster than the morality bully pulpit can change leaders and trigger meaningful action. Don't waste your energy, you just harden conservatives, encourage opposition, embolden enemies.

CATASTROPHIC ENDING? OR A NEW 'MEANING OF LIFE?'

Acceptance of this new reality is powerful. Not passively endurance, but acceptance that this is a time to focus of what really matters, preparing for a future, helping friends and family, living, rather than wasting time in ventures that have zero chance of succeeding.

Already "too late," says activist Bill McKibben. Physicist Stephen Hawking is even scarier, warns of a "catastrophic ending" for Planet Earth. Physicists generally share the long view, which capitalists cannot grasp in their fleeting world of daily closing prices, quarterly shareholders reports, annual bonuses.

Capitalists cannot understand Nobel Physicist Enrico Fermi's famous paradox: "Where is everybody? Why no signs of intelligence life in our universe?" Nowhere? Modern astrophysicist Adam Frank questions: "All planets hit a sustainability bottleneck, and none ever make it to the other side."

None? Why? Because every "civilization inevitably leads to catastrophic planetary change." Yes, we are the only intelligent life out there, our civilization is self-destructing, our home planet is dying.

You need a new consciousness. Living in the now, accepting the unstoppable. Old ways of living, working, investing, voting, loving, relating are dying, too.

Evangelicals also see what's ahead, in biblical warnings of End Days, Four Horsemen of the Apocalypse, the Book of Revelation and the passing of Heaven and Earth. Also seen powerful metaphors in brilliant theater Avatar, Star Trek, Interstellar, where humans leave their dying planet in search of new resources, a new home.

In this new normal, you live in a new consciousness where, in the words of the philosopher Pierre Teilhard de Chardin, a Jesuit priest: "You are not a human being in search of a spiritual experience. You are a spiritual being immersed in a human experience."

MASTER ALGORITHM IS A 'COLLECTIVE UNCONSCIOUS'

Anthropologist Jared Diamond, author of the classic *Collapse: How Societies Choose to Fail or Succeed,* tells how centuries ago two million people lived in the Mayan civilization. And like "so many societies the elite made decisions that were good for themselves in the short run and ruined themselves and societies in the long run."

As a result, their civilization collapsed "because of a combination of climate change, drought, water-management problems, soil erosion, deforestation." Their rulers "managed to insulate themselves from the consequences of their actions ... kings didn't recognize that they were making a mess ... until it was too late."

Yes, history is repeating again. No wonder Stephen Hawking sees a "catastrophic ending" ahead for our dying planet. Why? Because all civilizations eventually hit a sustainability bottleneck, and none ever make it to the other side. They all self-destruct. That's the new normal.

The good news is anyone can live in peace and prosperity on a dying planet. Acceptance is a choice, a new consciousness, new way of living, totally aware you are a spiritual being having human experiences. As Eckhart Tolle put it, a "New Earth arises as more and more people discover their main purpose in life is to bring the light of consciousness into this world and so use whatever they do as a vehicle for consciousness." And in the process, become part the transformation of Carl Jung's "collective unconscious" into a new global "collective consciousness."

8.10.15

ZOMBIE
APOCALYPSE

*"From a behavioral economics perspective, we are
fallible, easily confused, not that smart, and often irrational,
more like Homer Simpson than Superman. From this perspective,
it is rather depressing...we are standing in front of a really
difficult problem, because people are just designed
not to care about this…global warming will happen
in the future…to other people."*
Dan Ariely, Predictably Irrational

12 SELF-SABOTAGING WAYS WE DENY CLIMATE CHANGE

Yes, America's "sleepwalking into disaster" warns the U.N. secretary-general. Yes, zombies sleepwalking. Because "our brains are wired to ignore climate change," says George Marshall in his book, *Don't Even Think About It: Why Our Brains Are Wired to Ignore Climate Change.* And Nobel Economist Daniel Kahneman warns in a *NewScientist* interview: "I am deeply sorry, but I am deeply pessimistic. I really see no path to success on climate change."

The big takeaways here … the American brain is programmed to ignore, minimize and reject global warming … humans are sleepwalking, we will be unprepared for future disasters ... only 24% Americans believe climate is a problem ... even the world's leading brain researcher is pessimistic ... Kahneman sees "no path for success." Our human brain is sabotaging advance planning and future policies.

But why? What about the future of capitalism? Global economy? Planet Earth? If Kahneman is "deeply pessimistic," should the UN, USA, EU, leaders like Bill Gates, Bill McKibben, all other activists and policy makers just give up? Because whatever they do will be sabotaged by the "collective unconscious" of our climate science-denying brains?

PRICE OF REAL FREEDOM IS TOO RISKY, TOO COSTLY

Kahneman sees "climate change as a perfect trigger, a distant problem that requires sacrifices now to avoid uncertain losses far in the future." But we refuse to make these sacrifices, even see the problem. Why? Our brains focus instead on today's pocketbook issues, family, jobs, food, payday, a new smart phone, latest tech gadgets, hashtags, today stuff. Climate change is future stuff.

Americans love freedom. New cars. Need gas. Every day. We're focused on short-term issues. Is climate a "big issue?" Gallup polls says no. Only 24% of Americans think "climate change" is a big problem, near the bottom of 15 national problems polled. We'll worry about climate later.

But will later be too late to prepare? Yes. Environmentalists like Bill McKibben, Earth Institute's Jeff Sachs, and RiskyBusiness.org leaders Mayor Michael Bloomberg, former Treasury Secretary Hank Paulson and billionaire Tom Steyner are warning it may already be "too late."

Humans are unpredictable, contradictory, closet saboteurs. Big Oil creates nine million jobs, about $1 trillion in annual revenue. Vanguard Funds own over $15 billion of Exxon Mobil across 170 funds and 20 million shareholders. And our own government's just keeps giving Big Oil $4 billion a year in tax incentives.

'GOD'S INVISIBLE HAND' WIRED US TO SELF-DESTRUCT?

In Clive Hamilton's *Requiem for a Species: Why We Resist the Truth about Climate Change,* we're warned that Earth will soon "enter a chaotic era lasting thousands of years. Whether human beings would still be a force on the planet, or even survive ... one thing seems certain: there will be far fewer of us." Maybe even none. Remember, last time the planet went dark, the dinosaur species was wiped out. Never returned.

So today, a huge 76% of us, 235 million Americans, are admitting climate change is not a top priority for Americans. Get it? We think short-term. Yes, denial is wired into our brains, by God, by the Invisible Hand of capitalism, or some other grand power that reveals itself in many mysterious ways, like the brain research of Kahneman and other behavioral scientists. Here are 10 other ways the collective human brain, our conscious soul, is blocked from focusing on long-term problems:

1. NARCISSISM LIMITS FUTURE: ME-FIRST, NOW ... CLIMATE LATER
Experts at Earth Policy Institute and Worldwatch Institute agree with Gallup, asking: "Peak Production From a Planet in Distress: Can We Keep It Up?" No. Instead, the global economic system is "programmed to squeeze ever more resources from a planet in distress ... A mixture of population growth, consumerism, greed, and short-term

thinking by policy makers and business people seems to be inexorably driving human civilization toward a showdown with the planet's limits." But later is too late. No time to prepare.

2. FEAR 'DREAM' IS DYING, YET DEMAND MORE GROWTH, JOBS

Our brains are split: We want the good ol' American Dream. A world of hope, promises. Since 1776 we've banked on Adam Smith's vision of a "Land of Prosperity." Our brains keep telling us the glory days will return. A new Industrial Revolution. 3% GDP growth. Robert Gordon asks: "Is U.S. Economic Growth Over?" Inequality accelerates. The rich get richer. Is capitalism killing us? Gordon sees our GDP falling below 1% growth by 2100.

3. BELIEF IN THE IRRATIONAL MYTH OF PERPETUAL GROWTH

The economists' myth of perpetual growth is based on fairy-tale assumptions. We're wasting nonrenewable natural resources, will eventually destroy the planet. Traditional economists work for companies with short-term views. In a system that says: If we can't grow this quarter, the long-term is irrelevant. Our mythology has us on a suicide mission. Our brains are saboteurs.

4. OUT-OF-CONTROL POPULATION: MORE BABIES, MORE RETIREES

Remember, *Scientific American* says global population growth is "the most overlooked and essential strategy for achieving long-term balance," yet, by 2050 global population will explode from 7 to 10 billion. Five years ago Bill Gates, Buffett, Soros, Rockefeller, Winfrey, Bloomberg and other billionaires met secretly, all agreed: World's biggest time-bomb? Overpopulation. Jeremy Grantham says we can't feed 10 billion. Gates says 8.3 billion people is the limit. Jeffery Sachs' Earth Institute warns that even five billion may be too many.

5. TABOOS BLINDING US FROM CONTROLLING POPULATION

In "The Last Taboo," *Mother Jones* editor Julia Whitty hit the nail on the head: "What unites the Vatican, lefties, conservatives and scientists in a conspiracy of silence? Population." Unfortunately, hot-button issues ignite powerful reactions from religious fundamentalists. Politicians, global leaders are silent. Even if runaway population is killing our world, by the time we wake up, it'll be too late to prep.

6. OUR BRAINS SELF-DESTRUCTING, WIDEN INEQUALITY GAP

Anthropologist Jared Diamond says public-health advances have "increased life spans in the Third World." Still, today about "80% of the world's population" survives on a few dollars a day. In "The Price of Inequality," Nobel economist Joseph Stiglitz warns, "there's less equality of opportunity in the U.S today than in almost any advanced industrial country." Since 2008, the Super Rich have captured 93% of all income growth. Capitalism has sold its soul to the devil.

7. SHORT-TERM THINKING HELPS SCIENCE DENIERS GET RICH

Even the new Secretary of State, Exxon Mobil's former $40 million-a-year CEO Rex Tillerson admits climate change is real. But he says it's just an "engineering problem" with "an engineering solution." Humans "spent our entire existence adapting." Sea-level rise? "We'll adapt!" Even with the U.N.'s 2,500 climate scientists 97% certainty climate change could wipe us off the planet, we unconsciously support the world's biggest climate-science deniers, Big Oil, the Koch Bros, as they spend millions buying votes of American politicians, we keep buying more cars, more gas.

8. BIG OIL HAS NO PUBLIC CONSCIENCE, KILLING PLANET

The world has "1.4 trillion barrels of oil, enough to last at least 200 years," says U.S. Chamber of Commerce CEO Tom Donohue, quoting Big Oil stats ... "2.7 quadrillion cubic feet of natural gas, enough to last 120 years ... 486 billion tons of coal, enough to last more than 450 years." Yes, 200 years of oil. Too bad it'll kill us in 50 years, says McKibben in Rolling Stone. Burning all that will increase carbon emissions, trigger suicide in 50 years.

9. PLANET RUNNING OUT OF FOOD, CAN'T FEED 10 BILLION

Jeremy Grantham's firm manages $124 billion. He warns that by 2050 there will be an "inevitable mismatch between finite resources and exponential population growth" with a "bubble-like explosion of prices for raw materials." Commodity shortages are becoming a "threat to the long-term viability of our species;" it is "impossible to feed 10 billion people."

10. TOO MUCH FAITH IN TECH SOLUTIONS TO "BIG PROBLEMS"

In Robert Gordon's provocative paper predicting declining economic growth, not only will America's GDP drop under 1% by 2100, the main reason is that emerging new technologies will never match the rate of GDP since the Industrial Revolution. Why? Silicon Valley's mega-optimism cannot translate into results, won't reverse the decline in America's economic growth.

Why? Silicon Valley is already failing America, warns Jason Pontin, editor-in-chief of *MIT Technology Review* in a provocative article, "Why We Can't Solve Big Problems." *The Review's* cover hammered home our myopia: "You Promised Me Mars Colonies," warned astronaut Buzz Aldrin, "Instead, I Got Facebook." Government stopped investing in big problems, like a moon landing. Capitalists took over. So did short-term thinking.

So ask yourself again, did the Invisible Hand of capitalism program the human brain to self-destruct? And take with it the human race, civilization, the planet? Wake up, clues are everywhere, not just in the work of Kahneman, Marshall, Hamilton, Sachs, Ban Ki-moon, McKibben, Bloomberg, Paulson, Steyner. But everywhere, all around you.

Why is the human brain wired to think short-term, ignore future threats, refusing to prepare for the risks, obvious ones that will ultimately self-destruct everything we love? Was a pre-programmed neurological "kill-switch" of in our DNA code? And what could trigger it today? Resources wars? Mass hunger? Pandemics? Global warming?
10.1.14

MORALS 'FOR SALE'

HIGHEST BIDDER!

"Without being fully aware of the shift, Americans have drifted from having a market economy to becoming a market society where almost everything is up for sale, a way of life where market values seep into almost every sphere of life and crowd out or corrode important values." Michael Sandel, What Money Can't Buy: The Moral Limits of Markets

YES, OUR MORALS ARE "FOR SALE" AND IT'S KILLING US

Yes, capitalism really *is* working ... for the *Forbes Global Billionaires!* Their ranks swelled from 322 in 2000 to 1,810 in 2016. In fact, just 67 billionaires control as much wealth as the poorest 3.5 billion half of the world. Plus, Credit Suisse Bank says by 2100 there will be 11 trillionaire families in the world. Meanwhile, the income of the average American worker has stagnated for over a generation.

But for the vast majority of the world, however, capitalism is a failure. Over a billion live on less than two dollars a day. In his *Capital in the Twenty-First Century*, economist Thomas Piketty warns, this inequality gap is toxic, dangerous. As global population explodes from 7 billion to 10 billion by 2050, food production will deteriorate. No wonder Pope Francis warns: "Inequality is the root of social ills," fueling more poverty, hunger, wars, revolutions.

For years we've been asking: Why does the capitalist brain blindly drive down this irrational path of self-destruction? We found someone who brilliantly explains why free market capitalism is hell-bent on destroying itself and the world along with it: Harvard philosopher Michael Sandel, author of the new classic, *What Money Can't Buy: The Moral Limits of Markets* and his earlier classic, *Justice: What's the Right Thing to Do?*

FROM 'MARKET ECONOMY' TO NEW 'MARKET SOCIETY'

For more than three decades Sandel's been teaching us why capitalism is undermining human morality ... and why we keep denying their irrational behavior. Why do we bargain away our moral soul? His classes number over a thousand. You can even take his course online free. He even summarized capitalism's takeover of America's conscience in "What Isn't for Sale?" in *The Atlantic*. Listen:

"Without being fully aware of the shift, Americans have drifted from having a market economy to becoming a market society ... where almost everything is up for sale ... a way of life where market values seep into almost every sphere of life and sometimes crowd out or corrode important values, nonmarket values." Here's a short summary:

"The years leading up to the financial crisis of 2008 were a heady time of market faith and deregulation — an era of market triumphalism," says Sandel. "The era began in the early 1980s, when Ronald Reagan and Margaret Thatcher proclaimed their conviction that markets, not government, held the key to prosperity and freedom." And in the 1990s with the "market-friendly liberalism of Bill Clinton and Tony Blair, who moderated but consolidated the faith that markets are the primary means for achieving the public good."

So today, "almost everything can be bought and sold." Today "markets, and market values, have come to govern our lives as never before. We did not arrive at this condition through any deliberate choice. It is almost as if it came upon us," says Sandel. "Market values were coming to play a greater and greater role in social life. Economics was becoming an imperial domain. Today, the logic of buying and selling no longer applies to material goods alone. It increasingly governs the whole of life."

"FREE-MARKET" MYTH CONTROLS YOUR MORALS

Yes, it's everywhere: "Markets to allocate health, education, public safety, national security, criminal justice, environmental protection, recreation, procreation, and other social goods unheard of 30 years ago. Today, we take them largely for granted."

- **PUBLIC SERVICES**: for-profit schools, hospitals, prisons ... outsourcing war to private contractors ... police forces by private guards "almost twice the number of public police officers" ... drug "companies aggressive marketing of prescription drugs directly to consumers, a practice ... prohibited in most other countries."

- **ADVERTISEMENTS**: in "public schools ... busses ... corridors ... cafeterias ... naming rights to parks and civic spaces ... blurred boundaries, within journalism, between news and advertising ... marketing of 'designer' eggs and sperm ... buying and selling ... the right to pollute ... campaign finance ... comes close to permitting buying and selling of elections."

- **INVESTING**: "The financial crisis did more than cast doubt on the ability of markets to allocate risk efficiently. It also prompted a widespread sense that markets have become detached from morals." Then comes the big question:

So what? "Why worry that we are moving toward a society in which everything is up for sale?"

- **CORRUPTION**: "Putting a price on the good things in life can corrupt them ... markets don't only allocate goods, they express and promote certain attitudes toward the goods being exchanged." Also "corrupt the meaning of citizenship. Economists often assume that markets ... do not affect the goods being exchanged. But this is untrue. Markets leave their mark."

- **INEQUALITY**: "Where everything is for sale, life is harder for those of modest means." If wealth just bought things, yachts, sports cars, and fancy vacations, inequalities wouldn't matter much. "But as money comes to buy more and more, the distribution of income and wealth looms larger."

Sandel warns that America's new capitalist mindset is crowding out "nonmarket values worth caring about. When we decide that certain goods may be bought and sold," they become "commodities, as instruments of profit and use." But "not all goods are properly valued in this way ... Slavery was appalling because it treated human beings as a commodity, to be bought and sold at auction," failing to "value human beings as persons, worthy of dignity and respect; it sees them as instruments of gain and objects of use."

MORALS NOW JUST ANOTHER "COMMODITY FOR SALE"

Nor do we permit "children to be bought and sold, no matter how difficult the process of adoption can be." The same with citizenship ... jury duty ... voting rights ... "we believe that civic duties are not private property but public responsibilities. To outsource them is to demean them, to value them in the wrong way." Many things should never be commodities.

Sandel's core message is simple: "The good things in life are degraded if turned into commodities. So, to decide where the market belongs, and where it should be kept at a distance, we have to decide how to value the goods in question — health, education, family life, nature, art, civic duties, and so on. These are moral and political questions, not merely economic ones."

Unfortunately, we never had that debate during the 30-year rise of "market triumphalism. As a result, without quite realizing it — without ever deciding to do so — we drifted from having a market economy to being a market society." And "the difference is this: A market economy is a tool ... for organizing productive activity. A market society is a way of life in which market values seep into every aspect of human endeavor. It's a place where social relations are made over in the image of the market." The difference is profound.

Not only did the debate never happen. It may never. Why? Because now politicians aren't up to debating values, so are pushing us past the point of no return. Today's "political argument consists mainly of shouting matches on cable television, partisan vitriol on talk radio, and ideological food fights on the floor of Congress," says Sandel. So "it's hard to imagine a reasoned public debate about such controversial moral questions as the right way to value procreation, children, education, health, the environment, citizenship, and other goods."

CAPITALISM PAST MORAL "POINT-OF-NO-RETURN"

Can we change? "The appeal of using markets to put a price on public values, is that there's no judgment on the preferences they satisfy." Debate is unnecessary. Markets don't "ask whether some ways of valuing goods are higher, or worthier, than others. If someone is willing to pay for sex, or a kidney ... the only question the economist asks is 'How much?' Markets ... don't discriminate between worthy preferences and unworthy ones."

This much is guaranteed: Capitalism is eliminating moral values. As Sandel puts it: "Each party to a deal decides for him- or herself what value to place on the things being exchanged. This nonjudgmental stance toward values lies at the heart of market reasoning, and explains much of its appeal." But unfortunately, market capitalism has also "exacted a heavy price ... drained public discourse of moral and civic energy."

In the Reagan era S&L scandals a generation ago, over 1,000 bankers went to jail as felons. Today politicians protect crooked bankers. No Wall Street executives prosecuted for fraud after 2008's massive losses. Now Congress simply chastises them before "clawing back" millions from their excessive pay. But no jail time.

The good professor is a great teacher, with one glaring flaw, he's too idealistic, too quixotic. You don't have to be a fatalist to know that without a catastrophic economic collapse, free-market capitalists —the world's 1,826 billionaires, Wall Street bankers and corporate CEOs, hedge managers, lobbyists and every other special interest getting rich off consumerism and the new "Market Society" — will never voluntarily surrender their control over the American political system. Rather, they will blindly continue down their self-destructive path with a bizarre conviction they are somehow "divinely" guided by the "Invisible Hand" of Ayn Rand's "mutant capitalism" and America's insatiable consumerism.

Meanwhile, we have no choice but to wait patiently till the collapse, anxiously aware that our bizarre political system will just keep degrading America's moral values, pricing, buying, selling, trading morals like commodities, because in the final analysis, everything has a price, and everyone has a price in our hot new exciting Market Society.
5.26.15

RISE & FALL

OF THE GREAT AMERICAN EMPIRE

"The challenges that face the United States are often represented as slow-burning…threats seem very remote … What if history is not cyclical and slow-moving but arrhythmic? What if history is at times almost stationary but also capable of accelerating suddenly, like a sports car? What if collapse does not arrive over a number of centuries but comes suddenly, like a thief in the night?" Niall Ferguson, Colossus: The Rise and Fall of The American Empire

---

THE COLLAPSE IS SWIFT, SILENT, CHAOTIC, CERTAIN!

Sudden or slow, civilizations inevitably collapse: "One of the disturbing facts of history is that so many civilizations collapse," warns anthropologist Jared Diamond in *Collapse: How Societies Choose to Fail or Succeed.* Many "civilizations share a sharp curve of decline. Indeed, a society's demise may begin only a decade or two after it reaches its peak population, wealth and power."

Harvard's Niall Ferguson, one of the world's leading financial historians, echoes Diamond's warning: "Imperial collapse may come much more suddenly than many historians imagine. A combination of fiscal deficits and military overstretch suggests that the United States may be the next empire on the precipice." Yes, we are on the edge of collapse.

PEAK POWER, WEALTH, POPULATION … SUDDEN CRASH

Dismiss this warning? Everything you learned, everything you believe and everything driving today's political leaders is based on a misleading, outdated theory of history. The American Empire sits at the edge of a dangerous precipice, at risk of a sudden, rapid collapse.

Ferguson is brilliant, prolific and a contrarian. His works include the recent *Colossus: The Rise and Fall of The American Empire; Ascent of Money: A Financial History of the World; The Cash Nexus: Money and Power in the Modern World;* and *The War of the World,* a survey of the "savagery of the 20th century" where he highlights a profound "paradox that, though the 20th century was 'so bloody,' it was also 'a time of unparalleled progress.'"

Why? Throughout history imperial leaders inevitably emerge and drive their nations into wars for greater glory and "economic progress," which inevitably leads their nation into collapse suddenly and swiftly, into a demise that "may begin only a decade or two after it reaches its peak population, wealth and power."

You'll find Ferguson's latest work, "Collapse and Complexity: Empires on the Edge of Chaos," in *Foreign Affairs,* the journal of the *Council of Foreign Relations,* a nonpartisan think tank. His message negates all the happy talk in today's news, about economic recovery and new bull markets, about "hope," about a return to "American greatness" –from Washington politicians and Wall Street bankers.

EMPIRES COLLAPSE ... FIVE STAGES ACROSS HISTORY

Ferguson opens with a fascinating metaphor: "There is no better illustration of the life cycle of a great power than "The Course of Empire," a series of five paintings by Thomas Cole that hangs in the New York Historical Society. Cole was a founder of the Hudson River School and one of the pioneers of nineteenth-century American landscape painting; in The Course of Empire,' he beautifully captured a theory of imperial rise and fall to which most people remain in thrall to this day. Each of the five imagined scenes depicts the mouth of a great river beneath a rocky outcrop."

If you're unable to see them at the Historical Society, they're all reproduced in *Foreign Affairs,* all underscoring Ferguson's warnings that the "American Empire on the precipice," near collapse.

FIRST STAGE: THE 'SAVAGE STAGE' BEFORE EMPIRE

"In the first, 'The Savage Stage,' a lush wilderness is populated by a handful of hunter-gatherers eking out a primitive existence at the break of a stormy dawn." Imagine our history from Columbus' discovery of America in 1492 on through four more centuries as we savagely expanded across the continent.

SECOND STAGE: 'ARCADIAN OR PASTORAL STAGE'

As the American Empire flourishes in "the second picture, The Arcadian or Pastoral Stage, is of an agrarian idyll: the inhabitants have cleared the trees, planted fields, and built an elegant Greek temple." The temple may seem out of place. However, Cole's paintings were done in 1833-1836, not long after Thomas Jefferson built the University of Virginia using both classical Greek and Roman revival architecture.

As Ferguson continues this tour you sense you're actually inside the New York Historical Society, visually reminded of how history's great cycles do indeed repeat over and over. You are also reminded of one of history's great tragic ironies—that all nations fail to learn the lessons of history, that all nations and their leaders fall prey to their own narcissistic hubris and that all eventually collapse from within.

STAGE THREE: CONSUMMATION OF AMERICAN EMPIRE

"The third and largest of the paintings is 'The Consummation of Empire.' Now, the landscape is covered by a magnificent marble entrepôt, and the contented farmer-philosophers of the previous tableau have been replaced by a throng of opulently clad merchants, proconsuls and citizen-consumers. It is midday in the life cycle."

'The Consummation of Empire' focuses us on Ferguson's core message: At the very peak of their power, affluence and glory, leaders arise, run amok with imperial visions and sabotage themselves, their people and their nation. They have it all. But more-is-never-enough as greed, arrogance and a thirst for power consume them. Back in the early days of the Iraq war, Kevin Phillips, political historian and former Nixon strategist, also captured this inevitable tendency in *Wealth and Democracy:*

"Most great nations, at the peak of their economic power, become arrogant and wage great world wars at great cost, wasting vast resources, taking on huge debt, and ultimately burning themselves out." You even sense the "Consummation" of the American Empire occurred with the leadership handoff from Bill Clinton to George W. Bush.

Unfortunately, that peak may be behind us. Look back at the past generation: Reagan, Clinton, Bush, Henry Paulson, Ben Bernanke, Sarah Palin, Barack Obama, Mitt Romney and all future American leaders are merely playing their parts in this greatest of all historical dramas, repeating but never fully grasping the lessons of history in their insatiable drive for "economic progress," to recapture former glory ... while unwittingly pushing our empire to the edge, into collapse.

STAGE FOUR: DESTRUCTION OF ONCE-GREAT EMPIRES

Then comes 'The Destruction of Empire,' the fourth stage in Ferguson's grand drama about the life-cycle of all empires. In 'Destruction' we see "the city is ablaze, its citizens fleeing an invading horde that rapes and pillages beneath a brooding evening sky." Elsewhere in "The War of the World," Ferguson described the 20th century as "the bloodiest in history, one hundred years of butchery." Today's high-tech relentless news cycle, suggests that our 21st century world is a far bloodier return to savagery.

At this point, investors are asking themselves: How can I prepare for the destruction and collapse of the American Empire? Unfortunately, there is no solution in the Cole-Ferguson scenario, only an acceptance of fate, of destiny, of history's inevitable stages gracing the walls of the New York Historical Society.

But there may be one solution, from *Wealth, War and Wisdom* by hedge fund manager Barton Biggs, Morgan Stanley's former chief global strategist who warns us of the "possibility of a breakdown of the civilized infrastructure," urging us to buy a farm in the mountains. "Your safe haven must be self-sufficient and capable of growing some kind of food ... well-stocked with seed, fertilizer, canned food, wine, medicine, clothes, etc. Think Swiss Family Robinson." And when the rebels come looting, fire "a few rounds over the approaching brigands' heads."

STAGE FIVE: DESOLATION AFTER EMPIRE DISAPPEARS

"Finally, the moon rises over Desolation" in Cole's fifth painting. Ferguson: "There is not a living soul to be seen, only a few decaying columns and colonnades overgrown by briars and ivy." No attacking armies. No waste-collecting robots. No rockets shuttling a new Mars colony. The good news is the Earth may naturally regenerate itself without humans. We saw that possibility in Alan Weisman's brilliant *The World Without Us,* as steel buildings decayed over time. Even microbes will eventually eat indestructible plastics, and perhaps after many eons pass, a New Earth may emerge in all its glory, with new species, a new Garden of Eden.

DARK NIGHT, STOLEN SPORTS CAR, ACCELERATING, CHAOS, SUDDEN CRASH!

In a *Los Angeles Times* column, Ferguson asks: "America, a Fragile Empire: Here today, gone tomorrow, could the United States fall that fast?" And his answer is both emphatic and disturbing: "For centuries, historians, political theorists, anthropologists and the public have tended to think about the political process in seasonal, cyclical terms ... we discern a rhythm to history. Great powers, like great men, are born, rise, reign and then gradually wane. No matter whether civilizations decline culturally, economically or ecologically, their downfalls are protracted."

We are deceiving ourselves, convinced "the challenges that face the United States are often represented as slow-burning ... threats seem very remote." But "what if history is not cyclical and slow-moving but arrhythmic?" asks Ferguson.

What if history is "at times almost stationary but also capable of accelerating suddenly, like a sports car? What if collapse does not arrive over a number of centuries but comes suddenly, like a thief in the night?" Yes, in one of those catastrophic events Nassim Taleb calls a "Black Swan," a totally unpredictable and improbable event with "massive consequences."

Ferguson's final message about America's destiny comes from *Foreign Affairs:* "Conceived in the mid-1830s, Cole's great five-part painting has a clear message: all empires, no matter how magnificent, are condemned to decline and fall." Throughout history, empires function "in apparent equilibrium for some unknowable period. And then, quite abruptly, collapse. A sleeping driver can be all it takes to go over the edge of chaos."

3.9.10

THE GREAT
ESCAPE!

*"Mankind is in danger of destroying ourselves by
our greed and stupidity. Life on Earth is at an ever-increasing
risk of being wiped out by a disaster, such as a sudden
nuclear war, a genetically engineered virus, or other dangers.
The human race has no future if it doesn't
go to space." Stephen Hawking*

10 BILLION LEFT BEHIND WITH EARTH'S "BIG PROBLEMS!"

NASA loved it. *The Martian* with Matt Damon tending his vegetable garden. What a fabulous PR shot, a close-up that makes you forget the famed 1990 shot of Earth from The Voyager. Carl Sagan called it the Pale Blue Dot. Forget all the fabulous Hubble Space Telescope shots of black holes, supernovas, distant galaxies exploding.

There was a human actual farming veggies on Mars: NASA, Elon Musk, Jeff Bezos, they all love it. And all of us watching through 3-D glasses made it a box office hit. Also, must have jacked up the pre-booking price of a one-way trip to Mars that includes an extraordinary opportunity to buy and own a new condo out there on Isaac Asimov's *Red Planet,* 34 million miles from Earth, and a lifelong obsession of space-dreamers like Musk.

Then more good news from NASA, another boost from the great Cosmic PR Machine: Suddenly, magically, just coincidentally NASA announced it found signs of flowing water on Mars, salty yet real. *Time* magazine was ecstatic: "Flowing water means new hope in the search for life." Yes, hope for new signs of life. And with that, as if by magic, Wall Street, Washington, Madison Avenue and Hollywood all seemed on board racing with this fab-u-lous Mars PR-Hype Machine.

Yes, *The Martian* film was certain to convince Congress to cough up a few billion of federal tax dollars to bankroll and launch the next "Matt Damon" NASA crew of seven into space, headed for Mars. And next? How about the plans of America's most notable, ambitious and innovative interstellar dreamers, Elon Musk, the genius behind Tesla automobiles, and Amazon's brilliant creator Jeff Bezos?

$60 TRILLION TO 'PROTECT' 10 BILLION LEFT BEHIND

For years Musk, Bezos and others have been chasing this dream, planning of launching and migrating a few million of earthlings into space. But unfortunately, the Musk/Bezos super-ambitious plans for a yet to-be-built Mars Colonial Transporter to take a hundred or so "explorers" at a time may be DOA for three big reasons: Seriously check the math, the logic, economics, public policy:

- **HIGHER PRIORITIES HERE ON PLANET-EARTH … TODAY!**
  Planet-Earth's environment is deteriorating so rapidly scientists estimate it'll take $60 trillion to reverse the trend, an impossibly steep price tag on a planet where total GDP of all national economies is only $75 trillion. *The Guardian* news already warns global warming a "ticking time bomb" that will "undermine the global financial system." Stephen Hawking predicts a catastrophic ending. And a University of Hawaii study and others warn that the planet is on a short fuse. Our planet's atmosphere is overheating so rapidly that temperatures will pass a worldwide irreversible point-of-no-return by 2047, while in a parallel economic trend, America's GDP collapses below one percent by 2100.

- **TOO LITTLE, TOO LATE FOR HUMAN BILLIONS LEFT BEHIND!**
  Even if Musk's SpaceX Mars Colonial Transporters were leaving every month carrying a payload bigger than new Saturn rockets, ferrying 100 people to Mars each trip, they can't realistically get many humans off our planet fast enough, maybe a century to exit a few million? By comparison, Earth's population is increasing at an astronomical rate exceeding 100,000 humans every single day, tens of millions annually, with Planet Earth's population now at 7.3 billion and growing steadily to peak above 10 billion by 2050.

- **BAD POLICY, NEW DISTRACTION, TAX PERK FOR THE RICH!**
  A few years ago, Musk told *Wired* magazine ago the price of booking a flight on SpaceX Mars Colonial Transporter would be $500,000. In short, the wealthy are their primary target for marketing this trip, likely get a tax write-off. They're also a bit nutso to want to relocate in an environment that's minus 60 degrees and unbreathable outside. *The Guardian* newspaper even ran a column on "The psychology of isolation, confinement and 24-hour Big Brother," a warning "those sent to live and die on the red planet face untold risk of mental illness" while confined to a small inner space of roughly 500 square feet forever.

The Martian story begins with a NASA team storm-threatened, forced into an emergency return launch leaving Damon on Mars, presumed dead. Unfortunately, Musk's SpaceX Transporters and Bezos' Blue Origin Teams may never even get there in the first place, never get to build a Mars Colony while our leaders are fighting to solve the above three obstacles prior to launch. Still, that certainly won't end this latest Don Quixote version of the "Impossible Dream." Why? Americans are all dreamers, searching for a better future, even if it bankrupts us.

So, expect the "Mars-or-Bust" sales-marketing blitz to accelerate, shifting into hyper-drive for years to come. Photo-ops of Matt and Jessica. Water. Potato patches. Who could wish for more! In fact, NASA, SpaceX and Bezos' Blue Origin, must all be in pre-launch mode, churning out more and more press materials hyping the benefits of space travel to the new Mars, photo-shop updates of Asimov's dark dreary Red Planet.

Still, they may recapture the thrill of 19th century adventurers boarding a wagon-train to the Wild West. The market's hot. A rival Dutch group, Mars One, says 80,000 have applied. Imagine sales brochures: "Welcome to the greatest adventure of a lifetime. Escape to the New Mars Colony. Travel aboard the luxury Mars Colony's newest Transporters for a couple years, as an explorer, then retire at the universe's hot new destination, the Famous Red Planet."

National Geographic is a bit more realistic. In *Mars: Inside the High-Risk, High-Stakes Race to the Red Planet,* they warn us: "How well Mars explorers would perform on arrival is uncertain ... Hazards of the trip include bone loss and brain damage ... If the trip doesn't kill you, living there might."

ONE-WAY TRIP TO NEW GULAG ... WORTH $500,000?

For Musk, Bezos and others, this is childhood dream fulfilled ... and whether there really are millions of humans ready to pony up $500,000 each to make a one-way Mars trip ... whether Musk can build and deliver new super-transporters hauling a hundred humans and supplies each trip ... and whether Musk can design innovative new rocket ships more powerful than the earlier low-orbit Saturns with their 260,000-pound payload built special for use by NASA with assists from giants like Boeing, IBM, Douglas and North American Aviation ... well, all that remains to be seen but they will never give up the dream, no matter the cost. And is the Mars job market any better than here?

Bottom line: Do Americans and American politicians really have the political will, the good ol' American spirit driving us to transport just a few million Americans out of 310 million total, cramped in a spacebus for a year across 34 million miles to an isolated future on a desolate, long-dead planet, in conditions worse than our own dying planet? Can we revive the good ol' American Dream for this new push to Mars? This isn't 19th century expanding America West, *that* wagon train sailed long ago.

IRRATIONAL RACE FROM DYING PLANET TO DEAD ONE!

As Dan Ariely writes in *Predictably Irrational,* it's all about behavioral economics: here's what American capitalism has really been about since initial launch back in 1776: "Adam Smith's version of invisible hand does not exist." Never did. Ariely makes clear the real "Invisible Hand" is not Wall Street's clever manipulations ... not the selfishness and collective greed of millions of individual capitalists acting selfishly and separately ... not even God's "Invisible Hand" with human affairs.

After the Mars Colony Transporters take off, the real "Invisible Hand" will still be right here on Planet-Earth, flawed humans still here manipulating the "predictably irrational" self-sabotaging behavior of ten billion humans left stranded behind, all struggling to survive on a rapidly and irreversibly over-heating planet that's heading for the same dark cold fate as Mars. Is that irrational?

When I first saw Spielberg's *Close Encounters of The Third Kind* I identified so much with Dreyfus I got extremely excited when he boarded the UFO to leave Planet-Earth. This time we'll just be waving a cheery goodbye on our mobile phones, to all those brave adventurers who think they're escaping to the "New Land of Freedom" on the SpaceX Mars Colonial Transporter, aware that those of us who chose to remain here, well, at least we know it's our free will choice … at least we "think it's free!

10.6.15

YIN=YANG!

GENDER REBALANCING

*"Patriarchy is crumbling.
We are reaching the end of 200,000 years of human history
and the beginning of a new era in which women, and
womanly skills are on the rise. A new matriarchy is emerging.
For the first time in history the global economy is
becoming a place where women are finding
more success than men." Jennifer Homans' review of
Hanna Rosin, The End of Men and Rise of Women*

PATRIARCHY COLLAPSING AFTER 200,000 YRS, MEN PANIC!

"Men are losing their grip," writes Jennifer Homans in her *New York Times* review of *The End of Men and Rise of Women,* Hanna Rosin's bestseller. "Patriarchy is crumbling. We are reaching 'the end of 200,000 years of human history and the beginning of a new era' in which women, and womanly skills and traits, are on the rise." Warning guys, this is the last gasp of your male-dominated patriarchy.

Yes, women will lead in the future. And men hate that, are trapped in the past, as the presidential campaign made clear. Men live in fear, driven by primitive tribal urges to hold onto ancient symbols of power, resist being controlled by women.

Like Don Quixote tilting at windmills. Even more than chasing hot market tips, today's men frantically search for past glory ... in purple pills for erectile dysfunction, low-testosterone, anti-aging cosmetics, magic solutions to belly fat ... playing fantasy football over a six-pack in their man-cave ... fighting for open-carry ... joining militias ... reliving the 19th century Wild Wild West ... hero worship icons like John Wayne ... yearning for more of the good old days when old-guy paternalism ruled.

Even today's conservative war on women, their obsessive drive to control women's reproductive rights, comes from deep in the collective conscious of the male brain, as men feel threatened by this historic power shift. Yes, "a new matriarchy is emerging" says Homans. "For the first time in history, the global economy is becoming a place where women are finding more success than men ... run by young, ambitious, capable women ... taking matters into their own hands."

WOMEN LEADERS MAKE MACHO MEN FEEL POWERLESS

But the real tragedy is that today's threatened male ego is also sabotaging America's economy and our status as a world power. How? By fighting a losing battle to hold onto the old vestiges of their ancient macho-dominated patriarchy, the weak male ego is making matters worse, manifest in the GOPs seeming endless, self-destructive, do-nothing partisan political wars that at their core simply confirm how patriarchy is crumbling. First, their eight-year long childish strategy opposing and bullying a black male president. Now they fear a woman could win the presidency, lead America till 2024. No wonder the ancient male brain is in panic survival mode.

Yes, history is changing, fast: There were only four women in my University of Virginia Law School graduating class. Now over half are women. Same with other professions. So, here's how we see the single biggest global trend that's defining the 21st century: How women are replacing men as leaders in America and around the world. Here's a summary of the seven elements of this rapidly emerging trend as identified in *Fortune, The New York Times, Time* magazine, money manager Jeremy Grantham and other sources:

1. FEMALE BRAIN WIRED AS A STRATEGIC THINKER

Money manager Jeremy Grantham says our male-dominated patriarchal culture has created "an army of left-brained immediate doers." Wall Street, Corporate America and Silicon Valley's social-media commandos all think short term, discounting to zero longer-term social costs, like climate change and resource depletion, while ignoring reality, that we're living on a planet incapable of feeding the 10 billion predicted within a generation. The short-term thinking brains of our male-dominated capitalist world (closing process, quarterly earnings, annual bonuses) are not psychologically wired to solve the world's bigger long-term problems. The female brain is better designed.

2. NEW ECONOMY JOBS FAVOR WOMEN, LEVEL PLAYING FIELD

Men raised in macho cultures with traditional values feel even more threatened as women gain equality and power. As Homans put it in the *Times:* "The end of men is really the end of a manufacturing-based economy." Six million lost jobs since 2000, mostly men, creating a vacuum. As a result, "a new matriarchy is emerging: For the first time in history, the global economy is becoming a place where women are finding more success than men ... run by young, ambitious, capable women ... taking matters into their own hands."

Forget politics, this is the "new service economy, which doesn't care about physical strength," demanding skills that "come easily to women." Our educational system is preparing a new generation of women leaders: "Today 50% more women get college degrees, so even if fewer women are at the top, they are beginning to dominate professions like accounting, financial management, optometry, dermatology, forensic pathology and veterinary practices."

3. MORE WOMEN GAINING POWER IN CORPORATE AMERICA

Last year *Fortune* magazine's list of "The 50 Most Powerful Women" revealed that when the list was launched in 1998 there were only two women CEOs of Fortune 500 companies. Today, there are 19 women CEOs, at giants like IBM, PepsiCo, Xerox, Kraft and DuPont. Plus "more women wield more power than at any point in history," including many "guiding the future of the global economy," like IMF Managing Director Christine Lagarde and Germany's Chancellor Angela Merkel.

4. MORE WOMEN ELECTED TO LEGISLATIVE POWER

Males believe they still have power, in running the federal government. That delusion is disappearing, threatening male politicians as well as male carpenters. The National Foundation for Women Legislators says: "The greatest rising force in American politics today is not a political party, nor is it the lobbying community, it is women." In the early 1980s "women held a mere 10% of all state legislative seats in the country, today they hold 24% of 7,382 seats nationwide. Currently 17 women serve in the U.S. Senate and 73 serve in the U.S. House of Representatives." Plus, six women are state governors. By 2050, women will be the power majority.

5. MORE WOMEN IN GOVERNMENTS WORLDWIDE

Yes, the trend is sweeping the world. In "The Case for Optimism" about our future, Bill Clinton's *Time* magazine feature a couple years ago, he said the "world is getting better all the time," citing five ways, including technology, health care, green energy. His fourth key: "Women Rule." Worldwide, women now make up 20% of elected legislators, almost double 15 years ago: "This is good news, not only for the individuals themselves but also for entire societies." Why? "It's been proven that women tend to reinvest economic gains back into their families and communities more than men do." Men tend to narcissism. Yes, women not only think different from men, they think better, strategic, long-term.

6. WOMEN ARE NEW LEADERS IN 'FIGHT FOR THE FUTURE'

While so many men still resist, others are teaming up with women, equals working together. Here is Clinton's fifth reason for optimism: "Justice, The Fight for the Future." Until recently "the future has never had a big enough constituency." But things are changing, rapidly, because the survival of the planet requires new thinking, new strategies. Women get it. They are taking the lead. Clinton says we must "create a whole different mind-set. We are in a pitched battle between the present array of resources and attitudes and the future struggling to be born."

7. MEN DEFEND PATRIARCHY, SABOTAGE OWN FUTURE

This cultural war reveals how men are, unfortunately, their own worst enemy, desperately trying to hold onto power, and sabotaging their future. Look beyond the so-called "war on women" rhetoric in the political arena where men fight to control women and women's issues. Statistically they're out of touch with most Americans. Fighting a losing battle.

But look deeper into the brains of these male politicians. They are frightened little boys who feel threatened at a deep subconscious gut level. So, they react, double down, fight harder to go back to an old familiar power structure where Old Guys Rule, controlling women's reproductive rights and more.

GENDER POWER RISING STRONG, FAST, REBALANCING

No, we can't go back. Why? Because a huge cultural tidal wave is sweeping both men and women in its historic path. Yes, "men are losing their grip ... patriarchy is crumbling ... we are reaching 'the end of 200,000 years of human history ... the beginning of a new era' in which women, and womanly skills and traits, are on the rise." It is time men wake up to the power shift. No one can stop this historic shift, for the new economy just keeps empowering more women, preparing them for the future. Wake up guys ... before it's too late for you ... for America ... for the world.
9.2.14

STAR-CROSSED
PLANET∗EARTH

"In The Origin of Species, Darwin noted that seasons of extreme cold or drought were effective checks of species numbers. The process of change is not always subtle or gentle. Each of the 'big five' mass extinctions in the fossil record of life on earth the past 540 million years was accompanied by an environmental disruption. During each of these events, between 50% and 90% of all species perished." Scientific American

PLANET IRREVERSIBLY OVERHEATS, LAUNCHES TIME-CAPSULE

One very special *Star Trek: The Next Generation* episode haunts me, critics call it the best ever. From Stardate 45944.1, "The Inner Light" gives us a brief look into the star-crossed future of two civilizations. One boldly exploring new worlds. The other leaving behind a time capsule, a space probe, a brief snapshot of its history and final days. The ending of a planet and its civilization.

"The Inner Light" is a powerful metaphor for our own planet, in the near future, perhaps in 2047 the time of much anticipated Singularity when human intelligence and artificial intelligence machines merge into a super-intelligence that can save the world, or destroy it as in *The Terminator*.

The facts: The U.S.S. Enterprise is on a research mission, completing a magnetic survey of the Parvenium system when it encounters a probe floating in space. They bring it onboard. Suddenly a telepathic energy bolt knocks Capt. Jean-Luc Picard unconscious, he collapses.

He wakes up on a strange planet. Dazed, recovering from a fever, he cannot recognize his wife. She calls him "Kamin." His friends are concerned, he was delusional for some time. He mumbles on about being a starship captain.

Time passes. He gradually adapts to this new reality on this far-off world. Memories of his prior life slowly fade. He falls in love with his wife again, raises a family, his children give them grandchildren. He lives the quiet, peaceful life he never imagined on the other planet and in all his space travels.

The planet's natural resources gradually disappear ... temperatures rise, overheating all life ... water gets scarce .... desert lands replace forests and rich farmlands ... food supplies are depleted ... Kamin's family is living on a dying planet ... in the end, we see him standing alone ... a wide-brimmed straw hat shielding his eyes from the blinding sun ... watching the launch of a rocket in the distance ... liftoff ... soaring high into the clouds ... contrails disappearing into the heavens ... carrying the final record of a great civilization on a once-rich planet.

METAPHOR FOR A FAST OVER-HEATING PLANET-EARTH

Suddenly the probe powers off. Picard wakes up on the floor of the Enterprise Bridge. Like all time-travelers, only a few minutes had passed in this dimension ... Back in command ... Engines power up ... They accelerate to warp ... continuing on their mission, boldly going where no one has gone before.

Picard is left with bitter-sweet memories of a long, simpler life on a peaceful planet that vanished thousands of years earlier. Later, alone in his quarters, Picard begins playing the flute retrieved from within the drifting space probe. He knows how to play a familiar tune from that other life deep in his memory. A haunting melody fills the entire starship ... time and space ... fade to black.

We live with 7.3 billion other humans on our home planet today. By 2050 the United Nations estimates our population will rocket to 10 billion, all competing with four hundred million Americans for ever-scarcer resources, on an overheating planet that would be familiar to Picard.

Yes, huge odds against us, with the rest of the world outnumbering us 22-to-1, all demanding a bigger share of a better life. Every nation, every society, everyone fighting for their own personal version of the American Dream, in an unsustainable lifestyle war that will require the impossible resources of not one but six planets.

We face an impossible quandary in a world where harsh demographics of population growth, *the bubble of all bubbles,* becomes the force driving all other bubbles, economic, agricultural, political, cultural. The ultimate force driving us in an accelerating trajectory into an unsustainable reality on a planet that can never feed 10 billion people.

CAPITALISM ON WINNING STREAK, CLIMATE KEEPS LOSING

In one generation Planet-Earth will be incapable of feeding 10 billion people. Many already own over a million cars demanding on an endless supply of fossil fuels. This dilemma is clear in Naomi Klein's new classic, *This Changes Everything: Capitalism vs. the Climate:* "There's something fundamentally wrong with the way we're organizing our economy and thinking about our place on the planet." Yes, capitalism is destroying our planet's climate. And to paraphrase Einstein, we can't fix our capitalism problem with the capitalist's solutions.

And yet, capitalists believe that capitalism is the solution to everything. Stalemate. Only worldwide revolutions shift this paradigm, says Klein. Which are likely to come, but too late.

We're in denial: We get wake-up calls from many thought-leaders like Klein. But our brains are wired to ignore them, clouded by over-optimism. Humans talk a good game;

our leaders may sign well-intentioned agreements to reduce carbon emissions. But as Gallup polls confirm: Global warming simply is not a major national issue for 76% of Americans.

None of this is new: America had a huge wake-up call well over a decade ago at the launch of the Iraq War. In the "Pentagon's Weather Nightmare" *Fortune* analyzed a declassified George Bush Defense Department report on global warming and climate change, one that echoes the Capt. Picard's life as Kamin.

PENTAGON WARNS: "BY 2020 WARFARE WILL DEFINE LIFE ON EARTH"

Yes, our own Pentagon predicted global climate wars were dead ahead, coming long before 2050, some way they have already begun accelerating. Listen to the warriors: "The climate could change radically, and fast. That would be the mother of all national security issues ... massive droughts, turning farmland into dust bowls and forests to ashes ... by 2020 there is little doubt that something drastic is happening, an old pattern could emerge—with warfare defining human life."

That was America's war commanders on climate change back in 2003. They saw the future. Too many people, too few resources, too much competition, and the clock was ticking loudly. With 2020 dead ahead. Yes, we have been warned. So many, many times. Even by our war machine. But America just keeps digging, deeper and deeper, into our collective denial. We get wake-up calls. We ignore them. Nothing new, we prefer denying reality.

Are we "boldly going" anywhere? Should America launch its own time-capsule probe. Now? In 2020? Wait till 2050, although by then it may be too late to do anything about anything, let alone waste precious remaining resources on a time capsule probe and rocket ... especially since our scientists have discovered no evidence of any intelligent life anywhere in our vast universe the last 14 billion years.

10.14.14

The Fermi Paradox 1950
"Where is Everybody?"

Remember the "Fermi Paradox," it started in a series of questions during one of Nobel Laureate Enrico Fermi's visits to the Los Alamos National Labs in 1950. Fermi, known as the "architect of the nuclear age," was headed for lunch in the Fuller Lodge with fellow physicists Edward Teller, Herbert York and Emil Konopinski.

As usual, the conversation ventured wide across some deep topics—interstellar travel, flying sauces, aliens and extraterrestrial civilizations. Someone broke the ice with a joke about a *New Yorker* magazine cartoon. At the time trashcans were mysteriously disappearing from New York City streets.

The *New Yorker* cartoon showed "little green men" offloading the stolen trashcans from a flying saucer that had returned back to the aliens' home planet. Fermi was amused, then got serious, rattled off a haunting series of cosmic zingers, now known as the 'Fermi Paradox:'

> *"Could space ships travel faster than the speed of light? To reach us, they'd have to ... and if aliens really could travel faster than the speed of light, then where is everybody? ... Why do we see no signs of intelligence elsewhere in the universe? ... Humans could theoretically colonize the galaxy in a million years or so. ... And if they could, astronauts from older civilizations could do the same. ... So why haven't they come to Earth?"*

Why? Yes, where is everybody? Why have no aliens come to Earth? Why do we see no signs of intelligence ... *anywhere* in the universe? Hollywood's had answers, some inspiring, like *Contact, ET and Close Encounters of The Third Kind,* some threatening, like *Invasion of the Body Snatchers, District 19, and The Terminator.*

But still, where is everybody? Are there really no other intelligent beings, anywhere? No signs, nowhere, never? Ask yourself; Why are we all alone? Why? Alone is this vast universe? Why, do no planets ever get past the sustainability block?

Today we'd welcome aliens coming to haul off all the trashcans, the tons of waste and pollution that we're creating every day. Is it already too late? Yes, the collective consciousness of 33 MasterCoders is already warning us that sometime between 2020 and 2047, the momentum driving temperature increases will be irreversible, permanently locked in. Planet-Earth will have passed the point-of-no-return, with no aliens coming to haul off our trash, we are suffocating in a cosmic dump of our own making, one aliens won't want either.

Collectively the MasterCoders reveal the same answers layered one-on-top-of-another, collective answers to the four questions of the Master Algorithm our world must solve ... Who's killing life on Planet-Earth, and why? ... Can new technology save our world in time? ... If not, what's next for the human species after 2047? ... Finally, can we live collectively in peace and prosperity on a dying planet? And if not, what's the secret for you as an individual to living in peace on a dying planet?